# THREE PROGRAMS AND TEN CRITERIA

**EVALUATING AND IMPROVING ACQUISITION PROGRAM MANAGEMENT AND OVERSIGHT PROCESSES WITHIN THE DEPARTMENT OF DEFENSE**

Robert V. Johnson ◆ John Birkler

Prepared for the
Office of the Secretary of Defense

**National Defense Research Institute**

# RAND

Approved for public release; distribution unlimited

UC
263
.J59
1996

# PREFACE

Acquisition in the Department of Defense (DoD) is a major undertaking in which the defense agencies and the military departments expend significant funds to procure everything from research to development, to test and evaluation, to production, to operational support, and, finally, to obsolescence. The opportunities for problems to occur and the unique challenges posed in dealing with those problems in a high-technology environment require constant vigilance at all levels of management within DoD.

Problems in major defense acquisition programs, when accurately identified, can be a source of guidance for improving acquisition-management procedures. Synthesizing a set of lessons learned from an analysis of past problems, this report develops a framework for evaluating management practices in ongoing development and/or production programs. The framework then serves as the basis for reviewing and evaluating a top-priority development program in each Service: the Navy's F/A-18E/F aircraft, the Air Force's F-22 fighter aircraft, and the Army's RAH-66 Comanche armed reconnaissance helicopter.

To obtain an overall assessment of the management of acquisition programs, we compare those reviews with the criteria established in our framework. However, we do not directly compare the three programs, because it was not *our* intention to pick or choose the "best of the best." Each of the three programs is its Service's top priority. The program activities and current status of each program are well known to all levels of Service, DoD, and congressional organizations. There is no one "only" way to manage; thus, we have not attempted to directly evaluate the management approaches of one Service against those of others.

A successful acquisition program has multiple aspects—technical, political, and cultural—that work in unison. This report addresses the technical aspect—factors dealing with organizational structuring, reporting channels, standardization of information across parameters (cost, risk, time, etc.), design of management systems (e.g., the use of teams, review boards, program classifications), and the related actions that administrators take to carry out their work. The technical aspects are inherently more amenable to being described than are the political and cultural aspects. However, because political and cultural aspects influence how the technical aspects of program management are implemented, they affect both the relevance and effectiveness of the technical aspects of program management. The interactions across all three aspects are critical and ultimately affect program-management practices. A companion research effort is under way within RAND to discuss the political and cultural aspects.

> Paul Bracken and John Birkler, *An Alternative Framework for Managing Strategic Change in the Defense Acquisition Process*, Santa Monica, Calif.: RAND, forthcoming.

This current effort is part of a broader attempt by the Office of the Under Secretary of Defense for Acquisition and Technology (OUSD/A&T), Acquisition Program Integration, to improve the acquisition-management controls and oversight processes used in the defense acquisition system. During the course of our research, we identified and evaluated innovative approaches to program management. We hope this report encourages consideration of those approaches and their use by other program offices. This report has been prepared for government and industry officials, as well as for members of Congress and their staff, who have an interest and an understanding of the DoD acquisition process, and who are or have been a part of that process.

The analysis was performed in the Acquisition and Technology Policy Center of RAND's National Defense Research Institute, a federally funded research and development center supported by the Office of the Secretary of Defense, the Joint Staff, and the defense agencies. It was funded by the Director, Acquisition Program Integration, OUSD/A&T. The research, data collection, and analysis were carried out from January through September 1995.

# CONTENTS

Preface . . . . . . . . . . . . . . . . . . . . . . . . . . . . . . . . . . . . . . . iii
Figures . . . . . . . . . . . . . . . . . . . . . . . . . . . . . . . . . . . . . . . xi
Tables . . . . . . . . . . . . . . . . . . . . . . . . . . . . . . . . . . . . . . . xiii
Summary . . . . . . . . . . . . . . . . . . . . . . . . . . . . . . . . . . . . . xv
Acknowledgments . . . . . . . . . . . . . . . . . . . . . . . . . . . . . . . xxiii
Acronyms and Abbreviations . . . . . . . . . . . . . . . . . . . . . . . xxv

Chapter One
    INTRODUCTION . . . . . . . . . . . . . . . . . . . . . . . . . . . . 1

Chapter Two
    A FRAMEWORK FOR EVALUATING ACQUISITION
        SYSTEMS MANAGEMENT . . . . . . . . . . . . . . . . . . . . 5
    Transition from Past Programs to Current Programs . . . . . 5
    Program-Assessment Matrix . . . . . . . . . . . . . . . . . . . . . 7

Chapter Three
    PROGRAM DESCRIPTIONS . . . . . . . . . . . . . . . . . . . . . 13
    Navy F/A-18E/F . . . . . . . . . . . . . . . . . . . . . . . . . . . . 13
    Air Force F-22 . . . . . . . . . . . . . . . . . . . . . . . . . . . . . 14
    Army RAH-66 . . . . . . . . . . . . . . . . . . . . . . . . . . . . . 15

Chapter Four
    PROGRAM MANAGEMENT BY THE SERVICES . . . . . . . . 17
    Composite Program Management . . . . . . . . . . . . . . . . . 17
    Clear Lines of Authority Have Been Established . . . . . . . . 20
        Navy F/A-18E/F . . . . . . . . . . . . . . . . . . . . . . . . . . . 21

| | |
|---|---|
| Air Force F-22 | 21 |
| Army RAH-66 | 23 |
| Communication Is Encouraged | 23 |
|     Navy F/A-18E/F | 24 |
|     Air Force F-22 | 24 |
|     Army RAH-66 | 25 |
| CPM, C/SCS, DAES, Etc., Are Used at Service Acquisition Executive Levels | 26 |
|     Navy F/A-18E/F | 29 |
|     Air Force F-22 | 32 |
|     Army RAH-66 | 35 |
| A Risk-Management Program/Process Is Used | 38 |
|     Navy F/A-18E/F | 38 |
|     Air Force F-22 | 39 |
|     Army RAH-66 | 43 |
| Requirements Are Controlled | 49 |
|     Navy F/A-18E/F | 49 |
|     Air Force F-22 | 50 |
|     Army RAH-66 | 50 |
| DPRO Support Has Been Firmly Established | 50 |
|     Navy F/A-18E/F | 51 |
|     Air Force F-22 | 52 |
|     Army RAH-66 | 52 |
| Incentives Are Apparent | 53 |
|     Navy F/A-18E/F | 54 |
|     Air Force F-22 | 54 |
|     Army RAH-66 | 55 |
| Funding Is Stable | 55 |
|     Navy F/A-18E/F | 56 |
|     Air Force F-22 | 57 |
|     Army RAH-66 | 58 |
| Management Team Is Selected for Credibility and Stability | 59 |
|     Navy F/A-18E/F | 61 |
|     Air Force F-22 | 61 |
|     Army RAH-66 | 62 |
| Security Promotes Management Involvement | 62 |
|     Navy F/A-18E/F | 63 |
|     Air Force F-22 | 63 |
|     Army RAH-66 | 63 |

Chapter Five
   SUMMARY OF OBSERVATIONS AND
      RECOMMENDATIONS...................... 65
   Summary of Observations ........................ 65
   Recommendations ............................. 66

Appendix: OTHER SPECIFIC PROGRAM/SERVICE
      INITIATIVES................................. 69

# FIGURES

| | | |
|---|---|---|
| 4.1. | Service Management Model | 20 |
| 4.2. | Empty-Weight Technical Performance Measure for the Navy's F/A-18E/F Program | 31 |
| 4.3. | Empty-Weight (Less Engines) Technical Performance Measure for the Air Force F-22 Program | 33 |
| 4.4. | Empty-Weight Technical Performance Measure for the Army RAH-66 Program | 37 |
| 4.5. | F/A-18E/F Risk Management Identifies Five Levels of Uncertainty and Five Levels of Consequences | 40 |
| 4.6. | Risk Assessment of the E&MD Phase for the Navy F/A-18E/F Program | 41 |
| 4.7. | Air-Vehicle Integrated Master Schedule Performance to First Flight | 43 |
| 4.8. | Risk-Assessment Flow for the Army RAH-66 Program | 47 |
| 4.9. | Risk-Assessment Summary for the Army RAH-66 Program | 48 |
| 4.10. | F/A-18E/F E&MD Budget Chronology | 56 |
| 4.11. | F-22 Cumulative OSD and Congressional Funding Cuts | 58 |
| 4.12. | RAH-66 RDT&E Funding Changes | 60 |
| A.1. | F-22 SPO Organization Changes | 72 |
| A.2. | LMAS F-22 Organization Changes | 73 |
| A.3. | SPO/LMAS Organizational Compatibility | 74 |

# TABLES

| | | |
|---|---|---:|
| S.1. | Program-Assessment Matrix | xviii |
| S.2. | Composite Management Assessment | xx |
| 2.1. | Program-Assessment Matrix | 9 |
| 4.1. | Composite Management Assessment | 19 |
| 4.2. | F/A-18E/F Reporting Requirements and Related Activities | 25 |
| 4.3. | F-22 Reporting Requirements and Related Activities | 26 |
| 4.4. | RAH-66 Reporting Requirements and Related Activities | 27 |
| 4.5. | Use of CPM/DAES | 28 |
| 4.6. | Initial Air-Vehicle PRR Assessment | 44 |

# SUMMARY

During the past 15 to 20 years, many acquisition programs have encountered technical shortfalls, schedule slippage, and cost growth. Most such problems occurred in the development phase of the acquisition process. The significant investment in these programs and the potential for cost growth in these programs are concerns to all management levels within the Department of Defense (DoD), the Executive Branch, and the Legislative Branch. Over the past four decades, blue-ribbon panels, special study groups, and other management reviews have developed strategies aimed at improving the acquisition process. Such reviews originate when the administration changes and a new Secretary of Defense takes office.

For example, in spring 1989, the Defense Management Review (DMR) was chartered by the Secretary of Defense. It resulted in the establishment of a shorter, more direct chain of command between the Program Manager and the Service Acquisition Executive. For each of the Services, this new chain of command represented a major change in program-management reporting. As is to be expected, it took time for this new process to be understood completely and to become fully institutionalized.

Problems and successes in major defense acquisition programs, when accurately identified, can be a source of guidance for improving acquisition-management procedures. This report describes and qualitatively evaluates acquisition-management procedures in three aircraft development programs: the Navy's F/A-18E/F aircraft, the Air Force's F-22 fighter aircraft, and the Army's RAH-66 Comanche armed reconnaissance helicopter. The analysis is based on a framework of criteria derived from a review of lessons learned and problems identified in the acquisition management of prior major DoD programs.

The framework we developed consists of the following criteria:

- Lines of authority have been established and are clear. Defense Management Review issues and/or problems must not cause confusion, bickering, or a diminution of Program Manager (PM) responsibility and accountability.
- Communication is open (no secrets—all information is divulged; using all media and avenues, e.g., e-mail, written, verbal) and continuous at and between all levels of authority.
- Cost/Schedule Control System, Cost Performance Measurement, and other management reports are used as indicators of trends in program progress and for reporting program status.
- Risk-management techniques have been implemented.
- Program stability has been achieved through control of requirements.
- A strong government-industry support team (Program Office, functional support, Defense Plant Representative Offices [DPROs]) is present and has explicit mechanisms for coordinating responsibilities.
- Incentives for the Program Manager are adequate and positive.
- Funding is stable and adequate.
- Selection of best-qualified personnel for key acquisition-management positions is objective and regulated.
- Security requirements do not restrict adequate and sufficient management.

This set of criteria was developed by the authors from their past experience in DoD acquisition management and their judgment of aspects that must be present to afford a realistic opportunity for program success. While not guaranteeing success, the positive aspects of these criteria should form a baseline for good management. The ten criteria are not mutually exclusive; they intersect with and overlap the others to some extent. However, the matrix focuses on ten specific characteristics that should contribute to having a more successful program outcome than if any are excluded.

From this list, we formulated a program-assessment matrix, or assessment table, against which to evaluate existing program execution. For each item, we established specific attributes that would cause a program to be judged "good," "fair," or "poor" as listed in Table S.1.

After we reviewed the various acquisition and oversight aspects of the three programs, we formulated a composite assessment of all three programs, using the matrix in Table S.1. To prevent unfairly choosing the "best of the best," we present a composite rather than a side-by-side comparison. There is no one given way to manage well. How a program is managed is determined by technical, political, and cultural factors within each Service of the military. Important differences in these factors can exist, yet all three factors can be "working." Table S.2 presents the composite; information on each criterion for each of the three programs is discussed separately in Chapter Four.

We also provide, in Table S.2, our judgment of progress being made within the Department of Defense and the Services to address each criterion. For the most part, positive change (improvement) is occurring. Program instability was the one significant negative aspect we found in all three Services. Despite the progress DoD is making in creating the environment, in structuring the appropriate organizational framework, and in providing the tools and support necessary for managing these programs, significant program instability remains from constant budget perturbations in the Services, DoD, and the Congress.

Meeting or giving attention to the criteria in the framework does not guarantee success. However, it is the authors' judgment that meeting the criteria facilitates good outcomes and better management. To properly address the total subject of program management, political and cultural attributes will also need to be considered.

Table S.1

Program-Assessment Matrix

| Criteria | Good | Fair | Poor |
|---|---|---|---|
| Clear Lines of Authority Have Been Established<br>— Government Team<br>— Contractor Team | — DMR requirements have been implemented<br>— written charters show SAE-PEO-PM relationships<br>— structured program-execution plan is in place | — organization is in place<br>— some ambiguity exists in chain of command and responsibilities | — lines of responsibility and/or authority are unclear<br>— chain of command is confusing |
| Communication Is Encouraged<br>— Government Team<br>— Contractor Team<br>— Government/Contractor | — communication is completely open<br>— unique and innovative procedures are being followed<br>— communication—verbal and written—is continuous | — follows government policy<br>— follows (minimum) contractual requirements | — communication is limited or non-existent<br>— relations are strained<br>— all communication is in writing |
| CS2, CPM, DAES, Earned Value Are Used<br>— Government Team<br>— Government/Contractor | — active program is in place<br>— used by PM as management tool<br>— used by higher levels of organization | — contractual requirements are being met<br>— limited use of tools by management, at all levels | — contractual requirements are being met<br>— no use of tools by management |
| Risk-Management Program/Process Is Used | — active program is in place<br>— used by PM as management tool<br>— used by higher levels of organization<br>— a structured approach is taken | — some, or limited, structure to process | — no structure to process<br>— no risk-management process is in place |
| Requirements Are Controlled<br>— Operational<br>— Contract | — disciplined approach is in place<br>— requirements are in strict compliance with DMR<br>— changes in requirement occur infrequently | — approach defined<br>— user requirements are frequently changed | — DMR not followed<br>— requirements changed continuously |

Table S.1—continued

| Criteria | Good | Fair | Poor |
|---|---|---|---|
| DPRO Support Has Been Established | — DPRO is actively involved at all levels<br>— DPRO's inputs are used by management team | — program in place<br>— limited use of DPRO by management team | — not used |
| Incentives Are Apparent | — visible positive incentives can be recognized<br>— limited or no disincentives affect program execution | — positive incentives are weak or limited<br>— some disincentives are present and influence program execution | — no visible positive incentives<br>— disincentives are obvious and are detrimental to program |
| Funding Is Stable, and Control and Support (OSD, Congress, Service) Are Ensured | — funding is stable (and adequate)<br>— PM/PEO inputs are accepted | — funding is adequate<br>— defense of resources requires continuous PM/PEO actions | — funding is unstable<br>— funding changes frequently<br>— PM/PEO inputs are not accepted |
| Management Team Is Selected for Credibility and Stability<br>— Government<br>— Contractor | — PM is well accepted up/down chain of command<br>— PM selection follows an approved process<br>— a set plan for PM rotation is in place<br>— DAWIA is followed | — PM has limited program-management background<br>— there is no set plan for rotation based on milestone decisions<br>— PM questioned frequently on his/her actions | — PM changes frequently<br>— no basis established for PM selection<br>— PM's views are not accepted |
| Security Promotes Management Involvement | — security does not impede program execution and management | — influences management actions<br>— restricts some access to or oversight of program status | — greatly hinders adequate management<br>— overly restricts data review and program assessments |

NOTES: PEO = Program Executive Officer; CS2 = Cost/Schedule Control System; CPM = contract performance measurement; DAES = Defense Acquisition Executive Summary; DAWIA = Defense Acquisition Workforce Improvement Act.

## Table S.2

## Composite Management Assessment

| Criteria | Assessment | Rationale |
|---|---|---|
| Clear lines of authority have been established | Good to good-plus (Improving) | — DMR organizational relationship (SAE-PEO-PM) is apparent<br>— IPTs are used |
| Communication is encouraged | Good (Improving) | — Data are used and exchanged<br>— Various structured forums (written/verbal) |
| CS2, CPM, DAES, etc., are used | Good to good-plus (Improving) | — Used by management (government/contractor)<br>— Most IPTs are responsible for using these tools and do use them (cultural change)<br>— Real-time data are being used |
| Risk-management program/process is used | Fair-plus to good (Improving) | — Risk-mitigation techniques are actively used<br>— Different styles and emphasis at top levels [neutral] |
| Requirements are controlled | Good (Improving) | — DMR/PEO/PM have organizational control of requirements implementation[a]<br>— Requirements are stable<br>— Process for change is strict |
| DPRO support has been instituted and is firmly established | Good (Improving) | — DPRO is actively involved<br>— Participates at all levels<br>— Lead PI at DPROs for programs |
| Incentives are positive and apparent | Fair (Declining) | — Programs' being top priority of Services is motivation for success<br>— Schedules are budget-driven [disincentive] |
| Funding is stable; control and support are ensured | Fair-minus (Declining) | — Budgets have major instabilities<br>— Programs and/or contracts have experienced rephasings<br>— Support is lacking |
| Management team is selected for credibility and stability | Good (Improving) | — PMs have good backgrounds and experience<br>— DAWIA criteria are employed in selection<br>— PM-selection process is being formalized |
| Security promotes management involvement | Good (Improving) | — Security has no negative effects<br>— Controls are being reduced where possible |

[a]Users are not free to dictate changes at will. Block upgrades are made instead.

## RECOMMENDATIONS

From our examination and evaluation of the three top-priority acquisition programs of the Services (all three of which are in development phase), we make four recommendations that are important for DoD to ensure that the improvements in acquisition management toward which it is striving will continue:

- All three programs suffer from unstable funding. Program Managers must often spend time defending their programs rather than managing them. Therefore, DoD must take action to stabilize the budgets for executing major, high-priority development programs in the Services. With current fiscal constraints, it might not be possible to protect every major program, but it should be feasible to protect the budget of one or two programs in each Service so that, after formal milestone review and approval, the budget could be changed only by the Service secretary and the military Service Chief. Mid-level managers and staffs should be precluded from tinkering.

- Integrated product teams (IPTs) are becoming more prevalent on programs, and are being made so by both the government and industry, to integrate related functions in a teaming approach to program execution. Because IPTs contribute to better communication and consider all aspects of an acquisition program integrally, DoD should support the evolution and maturation of IPTs within DoD and within industry; learn what is being done within the IPTs, and what their experiences, good *and* bad, have been; and share this information. The IPT concept should be permitted to evolve, not be dictated from high levels of DoD as a policy directive detailing how to use them. Both government and industry would benefit.

- Whereas communication on most acquisition programs before the current administration was either mandated to be in writing or was predicated on saying something only if an individual also had a solution for a problem or for something perceived as negative, DoD is now supporting open communication of real-time status to all levels of program authorities. This support should be expanded, and the reporting of bad news should be encouraged by not taking immediate negative actions (such as an automatic budget reduction, creation of a special review team to investigate the issue, or a call for a major program review by the milestone decision authority). The Services, Program Executive Officers (PEOs), and Program Managers, should be given time to analyze the situation and develop alternatives and recovery paths.

- As a valuable extension of this research and to compare how industry and the commercial world manage and operate, it would be helpful for DoD to assess similar, major commercial programs, using the approach taken here to compare how industry and the commercial world manage and operate. Comparing styles of management, processes used, incentives, and oversight techniques could give DoD useful information and insights.

# ACKNOWLEDGMENTS

We would like to thank Irv Blickstein, Director, Acquisition Program Integration, for his guidance and forthright comments. In addition, the Army, Navy, and Air Force Service Acquisition Executives (SAEs), represented by their staffs at the SAE level; the Program Executive Officers (PEOs), and the three Program Managers and System Program Directors and their teams; the Defense Plant Representative Offices (DPROs) involved; and the RAH-66, F/A-18E/F, and F-22 contractor teams were helpful and willingly gave us their time and program data to assist in this effort.

# ACRONYMS AND ABBREVIATIONS

| | |
|---|---|
| AAE | Army Acquisition Executive |
| AAPERS | Army Acquisition Program Evaluation and Review System |
| ACAT | Acquisition category |
| A/C | Aircraft |
| ACC | Air Combat Command |
| ADM | Acquisition Decision Memorandum |
| AF | Air Force |
| AFB | Air Force Base |
| AFMC | Air Force Material Command |
| AFPD | Air Force Program Directive |
| AFSARC | Air Force System Acquisition Review Council |
| ALC | Air Logistics Center |
| AMAD | Aircraft-mounted accessory drive |
| ASA (RDA) | Assistant Secretary of the Army for Research, Development and Acquisition |
| ASAF (ACQ) | Assistant Secretary of the Air Force (Acquisition) |
| ASC | Aeronautical Systems Center |
| ASN (RDA) | Assistant Secretary of the Navy for Research, Development and Acquisition |
| ATCOM | Aviation and Troop Command |
| ATD | Achieved to date |
| AVRDEC | Aviation Research, Development and Engineering Center |
| AWOC | Acquisition Workforce Oversight Council |

| | |
|---|---|
| BH | Boeing Helicopters |
| BMA | Boeing Military Airplanes |
| BRAC | Base Realignment and Closure |
| CA | Contract award |
| CAO | Contract Administrative Officer |
| C2 | Command and control |
| CDR | Critical design review |
| CDRL | Contract Data Requirements List |
| CE | Current estimate |
| CECOM | Communication and Electronics Command |
| CEO | Chief Executive Officer |
| CFE | Contractor-furnished equipment |
| CNO | Chief of Naval Operations |
| COMNAVAIRSYSCOM | Commander, Naval Air Systems Command |
| CPM | Contract performance measurement |
| CPR | Cost performance report |
| C/SCS | Cost/Schedule Control System (referred to as CS2) |
| CS2 | Cost/Schedule Control System |
| DA | Department of the Army |
| DAB | Defense Acquisition Board |
| DAC | Designated Acquisition Commander |
| DAES | Defense Acquisition Executive Summary |
| DAWIA | Defense Acquisition Workforce Improvement Act |
| DCMC | Defense Contract Management Command |
| Dem/Val | Demonstration/validation |
| DLA | Defense Logistics Agency |
| DMR | Defense Management Review |
| DoD | Department of Defense |
| DODD | Department of Defense Directive |
| DODI | Department of Defense Instruction |
| DON | Department of the Navy |
| DPRO | Defense Plant Representative Office |
| DSMC | Defense Systems Management College |
| DTC | Design-to-cost |
| DTW | Design-to-weight |

| | |
|---|---|
| EAC | Estimate at completion |
| ECP | Engineering change proposal |
| E&MD | Engineering and Manufacturing Development |
| EOC | Early operational capability |
| ESG | Executive Steering Group |
| EW | Electronic warfare |
| FIRM | F-22 Information Resources Management |
| FMS | Foreign Military Sales |
| GDT | Government development team |
| GE | General Electric |
| GFE | Government-furnished equipment |
| GFM | Government-furnished materiel |
| ILS | Integrated logistics support |
| IMICS | Integrated management information and control system |
| IMP | Integrated master plan |
| IMS | Integrated master schedule |
| IPD | Integrated product development |
| IPT | Integrated product team |
| IR | Infrared |
| IWSM | Integrated Weapon System Manager |
| JDAM | Joint Direct Air Munition |
| JPO | Joint Program Office |
| KTR | Abbreviation for *contractor* |
| LASC | Lockheed Aeronautical Systems Company |
| LFWC | Lockheed Fort Worth Company |
| LHTEC | Light Helicopter Turbine Engine Company |
| LIMDIS | Limited distribution |
| LL | Lessons learned |
| LMAS | Lockheed Martin Aeronautical Systems |
| LMTAS | Lockheed Martin Tactical Aircraft Systems |
| LO | Low observable |
| LRIP | Low-rate initial production |
| MANPRINT | Manpower and personnel integration |
| MAPR | Monthly acquisition program review |
| MDA | McDonnell Douglas Aircraft |

| | |
|---|---|
| MEP | Mission equipment package |
| MIS | Management information system |
| MOA | Memorandum of Agreement |
| MPCD | Multi-purpose color display |
| MS | Milestone |
| MSC | Major subordinate command |
| M/TIS | Management/technical information system |
| MWG | Management working group |
| NAD | Northrop Aircraft Division |
| NAVAIR | Naval Air Systems Command |
| NAVCOMPT | Navy Comptroller |
| NGC | Northrop Grumman Corporation |
| NT | New technology |
| NVPS | Night-vision pilot system |
| ORD | Operational Requirements Document |
| OSD | Office of the Secretary of Defense |
| OUSD/A&T | Office of the Under Secretary of Defense for Acquisition and Technology |
| PAT | Process-action team |
| PCE | Production cost estimate |
| PD | Program Director |
| PDASN (RDA) | Principal Deputy Assistant Secretary of the Navy (Research, Development and Acquisition) |
| PDR | Preliminary design review |
| PDT | Product development team |
| PEM | Program element monitor |
| PEO | Program Executive Officer |
| PEO(T) | Program Executive Officer for Tactical Aircraft |
| PEO/TA | Program Executive Officer for Tactical and Airlift Programs |
| PFQ | Preliminary flight qualification |
| PI | Program integrator |
| PIA | Program independent analysis |
| PM | Program Manager |
| PMA | Program Manager Air |

| | |
|---|---|
| PMO | Program Manager's Office |
| POC | Point of contact |
| P3I | Pre-planned product improvement |
| PPBS | Planning, Programming and Budget System |
| PRO | Plant Representative Office |
| PRR | Production readiness review |
| P&W | Pratt and Whitney |
| QA | Quality assurance |
| RDT&E | Research, development, test and evaluation |
| RFP | Request for proposal |
| RPT | Report |
| RTM | Requirements Traceability Matrix |
| SA | Secretary of the Army; Sikorsky Aircraft |
| SAE | Service Acquisition Executive |
| SAF/AQ | Assistant Secretary of the Air Force (Acquisition) |
| SAF/AQC | Deputy Assistant Secretary of the Air Force (Contracting) of SAF/AQ |
| SAF/AQP | Director, Fighter, C2 and Weapons Programs of SAF/AQ |
| SAF/AQX | Deputy Assistant Secretary of the Air Force (Management Policy and Program Integration) of SAF/AQ |
| SAMP | Single Acquisition Management Plan |
| SAP | Special-access program |
| SAR | *Selected Acquisition Report* |
| SECAF | Secretary of the Air Force |
| SECNAVINST | Secretary of the Navy Instruction |
| SES | Senior Executive Service |
| SMM | System maturity matrix |
| SOW | Statement of Work |
| SPD | SPO Program Director |
| SPM | System Program Manager |
| SPO | System Program Office |
| SSEB | Source Selection Evaluation Board |
| SSM | System Support Manager |
| SSPO | Strategic System Programs Office (Navy) |

| | |
|---|---|
| SYSCOM | Systems Command |
| TA | Tactical aircraft |
| TAS | Target-acquisition system |
| TPM | Technical performance measure |
| TRADOC | Training and Doctrine Command |
| TSM | TRADOC Systems Manager |
| USA | United States Army |
| USAF | United States Air Force |
| USD/A&T | Under Secretary of Defense for Acquisition and Technology |
| USN | United States Navy |
| VMS | Velocity measurement system |
| VP | Vice President |
| VTC | Video teleconferencing center |
| WBS | Work Breakdown Structure |
| WPAFB | Wright-Patterson Air Force Base |

Chapter One
# INTRODUCTION

Acquisition in the Department of Defense (DoD) is a major undertaking in which the defense agencies and the military departments expend significant funds to procure everything from research to development, to test and evaluation, to production, to operational support, and, finally, to obsolescence. The opportunities for problems to occur and the unique challenges posed in dealing with those problems in a high-technology environment require constant vigilance at all levels of management within DoD.

During the past 15 to 20 years, many acquisition programs have encountered technical shortfalls, schedule slippage, and cost growth. Most such problems occurred in the development phase of the acquisition process. The significant investment in these programs and the potential for cost growth in these programs are concerns to all management levels within DoD, the Executive Branch, and the Legislative Branch. Over the past four decades, blue-ribbon panels, special study groups, and other management reviews have developed strategies aimed at improving the acquisition process. Such reviews originate when the administration changes and a new Secretary of Defense takes office.

For example, in the spring of 1989, the Defense Management Review (DMR) was chartered by the Secretary of Defense. It resulted in the establishment of a shorter, more direct chain of command between the Program Manager and the Service Acquisition Executive. For each of the Services, this new chain of command represented a major change in program-management reporting. As is to be expected, it took time for this new process to be fully understood and to become fully institutionalized.

Problems in major defense acquisition programs, when accurately identified, can be a source of guidance for improving acquisition-management procedures. This report presents the results of three case studies of major aircraft acquisition programs (all of which are in the development phase) to which we applied one set of criteria as a way of evaluating acquisition-management procedures in those programs. The criteria were developed from our examination of acquisition issues and from lessons learned from problems in earlier programs.

The objective of this work is to help improve acquisition-management controls and oversight processes used in the defense acquisition system. To accomplish this objective, we address the following question:

> *How are specific DoD developmental programs being managed, especially in light of the shortcomings that have been recognized and are generally believed to exist in program management of prior programs?*

Where the answer to this question discloses good ideas and the tools and processes that are achieving increases in communication and reductions in cost and schedule in one or more programs, we hope this report encourages consideration of those ideas, tools, and processes and their use by other Program Managers.

Addressing this question involved two tasks: first, to identify significant issues and problems and/or lessons learned from past programs, and to develop a framework—specifically, a program-assessment matrix—for reviewing management processes on current programs; second, to use the matrix to review three major DoD acquisition programs that are now in the development phase—the Navy's F/A-18E/F Engineering and Manufacturing Development Program, the Air Force's F-22 Engineering and Manufacturing Development Program, and the Army's RAH-66 Comanche Demonstration/Validation Prototype Program.

Each of these three programs was jointly selected for study, by RAND and the Office of the Under Secretary of Defense for Acquisition and

Technology (OUSD/A&T), Acquisition Program Integration Office, because it represents its Service's top-priority acquisition programs, was thought to be well managed, and was using innovative management techniques. In addition, all three are aerospace programs, which enabled the research team to focus on management approaches and techniques within a given technology sector instead of having to rationalize aspects that might differ across various technologies.

We had many discussions with personnel from the Service Acquisition Executives (SAEs) of the three Services, the Program Executive Officer (PEO) staffs, Defense Plant Representative Offices (DPROs), System Commands (SYSCOMs) and their equivalents, the Defense Systems Management College (DSMC), and the Program Offices (including both government and industry officials). We received briefings, obtained written documentation, and conducted a literature search to identify appropriate DoD and Service regulations, directives, and/or instructions. We visited air-vehicle prime contractors to see first hand how industry was managing these major defense programs.

The remainder of the report is organized as follows. Chapter Two presents the evaluation framework. Chapter Three gives a brief overview of the three programs. Chapter Four applies the ten criteria in the evaluation framework to each program separately and presents a composite program-management evaluation. Chapter Five summarizes observations and recommendations. The Appendix details important management tools and organizations mentioned briefly in the text, and describes specific examples of successful practices merely alluded to in the text.

Chapter Two

# A FRAMEWORK FOR EVALUATING ACQUISITION SYSTEMS MANAGEMENT

In this chapter, we summarize problems encountered in prior programs, present a list of key factors we believe to be important to successful program management, and, finally, develop a program-assessment template to evaluate these factors in the three subject programs.

## TRANSITION FROM PAST PROGRAMS TO CURRENT PROGRAMS

Over the past ten years, several significant changes have been implemented in the DoD acquisition arena, resulting in the initiation of organizational changes and processes that were intended to improve weapon systems acquisition management. As a first step in this analysis, we look at these changes and at the resulting current and future programs, to provide a basis for understanding that, when significant changes occur in the acquisition process, careful attention must be paid to their implementation procedures and phasing of implementation, and that added safeguards may be necessary to ensure that something, e.g., a confusing chain of command or uncertainty of personal authority or who is in charge, has not "fallen through the cracks." These changes were

- complete revision of the DoD 5000-series directives, instructions, and procedures.[1]
- shifting of acquisition responsibilities from military to civilian control.
- DMR implementation of the Program Executive Officer concept, with major program responsibility shifting away from the System Commands.
- transfer of Service Plant Representative Offices (PROs) to the central Defense Logistics Agency (DLA) and the new Defense Contract Management Command (DCMC) organization.
- establishment of new Service Acquisition Executives' responsibilities, along with major realignments of the acquisition organizations.
- strong emphasis on security implications of special-access programs (SAPs), and the desire to restrict access to people with management oversight as a way of protecting the security of advanced technologies.
- shifting to fixed-price contracts for relatively high-risk development programs, in the belief that the best management technique is an arms-length, hands-off approach that lets the contractors proceed without significant and continuous government-industry interface or oversight.
- influx of new leadership with different backgrounds and experiences at the Service secretariat and PEO levels.

None of the above actions, taken individually or collectively, can be cited as the cause of program problems. However, a significant number of changes were going on at the same time in various related areas. Many such changes were due to political and cultural factors that affected the technical dimensions of program management, in turn contributing to a lack of rigorous, disciplined, institutionalized

---

[1]DoD Directive 5000.1, "Defense Acquisition," February 23, 1991; DoD Instruction 5000.2, "Defense Acquisition Procedures," February 23, 1991; and DoD Manual 5000.2M, "Defense Acquisition Management Documents and Reports," February 23, 1991. (These documents have been revised again, but their publication date is March 15, 1996, which is subsequent to the material covered in this research effort.)

processes for managing programs and for communicating among government and industrial management officials. Collectively, the changes created uncertainty about what process or procedure changes were being made. Such uncertainty, in turn, led to uncertainty about how to act or react to changing conditions.

In making the transition from the review of prior programs to an examination of current programs, we believe it is necessary to break away from individual issues or problems and translate them into a framework for assessing program-management controls and the processes being used.

From our research on past programs, we identified many shortcomings in program management and the processes used to control programs. We used our backgrounds and general knowledge of past and ongoing DoD programs, both fixed-wing and rotary-wing, as well as items and issues identified in the public domain, to derive a list of factors for assessing or evaluating whether a DoD acquisition program is well managed and can be developed successfully. In most of the areas identified with respect to past programs, the DoD has taken significant steps to implement recommendations and to reinforce and/or revise current regulations, directives, and instructions, or to develop new ones.

## PROGRAM-ASSESSMENT MATRIX

The preceding observations, combined with our backgrounds and knowledge of past programs' problems and lessons learned, led us to develop a structured list of key factors to be considered in successfully managing a major defense acquisition program. The factors are as follows:

- Lines of authority have been established and are clear. Defense Management Review issues and/or problems must not cause confusion, bickering, or a diminution of Program Manager responsibility and accountability.

- Communication is open (no secrets—all information is divulged; using all media and avenues, e.g., e-mail, written, verbal) and continuous at and between all levels of authority.

8  Three Programs and Ten Criteria

- Cost/Schedule Control System (CS2), cost performance measurement (CPM), and other management reports are used as indicators of trends in program progress and for reporting program status.
- Risk-management techniques have been implemented.
- Program stability has been achieved through control of requirements.
- A strong government-industry support team (Program Office, functional support, Defense Plant Representative Offices) is present and has explicit mechanisms for coordinating responsibilities.
- Incentives for the Program Manager are adequate and positive.
- Funding is stable and adequate.
- Selection of best-qualified personnel for key acquisition management positions is objective and regulated.
- Security requirements do not restrict adequate and sufficient management.

This set of criteria was developed by the authors from their past experience in DoD acquisition management and their judgment of aspects that must be present to afford a realistic opportunity for program success. While not guaranteeing success, the positive aspects of these criteria should form a baseline for good management. The ten criteria are not mutually exclusive; they intersect with and overlap the others to some extent. However, the matrix focuses on ten specific characteristics that should contribute to having a more successful program outcome than if any are excluded.

From this list we developed the program-assessment template shown in Table 2.1. It provides the elements of a basic evaluation process for judging ten specific factors related to managing a major acquisition program. To arrive at this matrix and to rationalize our assessment criteria, we used our judgment, knowledge of past program failures, and some positive attributes accorded to successful programs.

## Table 2.1
### Program-Assessment Matrix

| Criteria | Good | Fair | Poor |
|---|---|---|---|
| Clear Lines of Authority Have Been Established<br>— Government Team<br>— Contractor Team | — DMR requirements have been implemented<br>— written charters show SAE-PEO-PM relationships<br>— structured program-execution plan is in place | — organization is in place<br>— some ambiguity exists in chain of command and responsibilities | — lines of responsibility and/or authority are unclear<br>— chain of command is confusing |
| Communication Is Encouraged<br>— Government Team<br>— Contractor Team<br>— Government/Contractor | — communication is completely open<br>— unique and innovative procedures are being followed<br>— communication—verbal and written—is continuous | — follows government policy<br>— follows (minimum) contractual requirements | — communication is limited or non-existent<br>— relations are strained<br>— all communication is in writing |
| CS2, CPM, DAES, Earned Value Are Used<br>— Government Team<br>— Government/Contractor | — active program is in place<br>— used by PM as management tool<br>— used by higher levels of organization | — contractual requirements are being met<br>— limited use of tools by management, at all levels | — contractual requirements are being met<br>— no use of tools by management |
| Risk-Management Program/Process Is Used | — active program is in place<br>— used by PM as management tool<br>— used by higher levels of organization<br>— a structured approach is taken | — some, or limited, structure to process | — no structure to process<br>— no risk-management process is in place |
| Requirements Are Controlled<br>— Operational<br>— Contract | — disciplined approach is in place<br>— requirements are in strict compliance with DMR<br>— changes in requirement occur infrequently | — approach defined<br>— user requirements are frequently changed | — DMR not followed<br>— requirements changed continuously |

## Table 2.1—continued

| Criteria | Good | Fair | Poor |
|---|---|---|---|
| DPRO Support Has Been Established | — DPRO is actively involved at all levels<br>— DPRO's inputs are used by management team | — program in place<br>— limited use of DPRO by management team | — not used |
| Incentives Are Apparent | — visible positive incentives can be recognized<br>— limited or no disincentives affect program execution | — positive incentives are weak or limited<br>— some disincentives are present and influence program execution | — no visible positive incentives<br>— disincentives are obvious and are detrimental to program |
| Funding Is Stable, and Control and Support (OSD, Congress, Service) Are Ensured | — funding is stable (and adequate)<br>— PM/PEO inputs are accepted | — funding is adequate<br>— defense of resources requires continuous PM/PEO actions | — funding is unstable<br>— funding changes frequently<br>— PM/PEO inputs are not accepted |
| Management Team Is Selected for Credibility and Stability<br>— Government<br>— Contractor | — PM is well accepted up/down chain of command<br>— PM selection follows an approved process<br>— a set plan for PM rotation is in place<br>— DAWIA is followed | — PM has limited program-management background<br>— there is no set plan for rotation based on milestone decisions<br>— PM questioned frequently on his/her actions | — PM changes frequently<br>— no basis established for PM selection<br>— PM's views are not accepted |
| Security Promotes Management Involvement | — security does not impede program execution and management | — influences management actions<br>— restricts some access to or oversight of program status | — greatly hinders adequate management<br>— overly restricts data review and program assessments |

NOTES: DAES = Defense Acquisition Executive Summary; DAWIA = Defense Acquisition Workforce Improvement Act.

Certainly, having completely open communication will not guarantee success; however, for government-to-government and government-to-contractor relations to be strained, non-existent, or relegated to written communication is obviously a "poor" way of doing business. Similarly, the other nine criteria we developed are, in our opinion, factors that would need to be present for assessing the implementation of the criteria as "good," "fair," or "poor."

We applied this matrix to three programs for systems currently in development. In Chapter Four, we describe each evaluation separately, then present a composite evaluation. We make recommendations for additional actions in Chapter Five. Results over time will tell whether the changes made and management approaches used in these three programs truly provide program results that meet performance, schedule, and cost constraints, by which all programs are evaluated. Meeting these criteria does not guarantee success. However, it is the authors' belief that addressing these criteria facilitates good outcomes and better management.

Chapter Three
# PROGRAM DESCRIPTIONS

The F/A-18E/F, the F-22, and the RAH-66 were jointly selected by RAND and the Director of Acquisition Program Integration of OUSD/A&T for this research because they represent their respective Service's top-priority acquisition programs, were thought to be well managed, and have been and are using innovative management techniques. In addition, all three are aerospace programs, which allowed the RAND research team to concentrate on program-management approaches and processes instead of having to rationalize differences that might be peculiar to a particular technology area.

## NAVY F/A-18E/F

The F/A-18E/F represents a major modification to its predecessor aircraft, which number over 1,100 produced for the Navy, Marine Corps, and Foreign Military Sales (FMS) customers.[1] The E/F will have higher-thrust engines, a 34-inch fuselage extension, 33 percent additional internal fuel, and two additional multimission weapon stations on the wing. It will have a longer range, increased payload (and bring-back capability), improved survivability, and future growth capability.

---

[1] U.S. Navy, "PMA-265 F/A-18E/F Program Update Briefing" (to RAND), Arlington, Va., January 26, 1995; McDonnell Douglas Aircraft, "MDA F/A-18E/F Program Overview Briefing" (to RAND), St. Louis, Mo., February 1, 1995.

The program officially started Engineering and Manufacturing Development (E&MD) with a successful Defense Acquisition Board (DAB) in May 1992.[2] It was approved at a cost of $4.88 billion (FY 90 dollars) and is congressionally capped at that amount. The E&MD phase is scheduled to take seven and a half years, and the first flight of one of seven test vehicles was scheduled for December 1995 (it actually flew at the end of November 1995). McDonnell-Douglas Aircraft (MDA) and General Electric (GE) are the two prime contractors; they report to the Navy through a Navy Program Manager Air (PMA-265) located at the Naval Air System Command (NAVAIRSYSCOM), Arlington, Virginia, and to the Program Executive Officer for Tactical Aircraft Programs. Northrop Grumman Corporation (NGC) and Hughes are key subcontractors. Both primes are performing on Navy (NAVAIR) cost-plus-incentive/award-fee contracts.

## AIR FORCE F-22

The F-22 is a fighter aircraft with a completely new design, technological advancements for which include stealth, supercruise, thrust vectoring, sensor fusion, integrated advanced avionics, and composite structure to provide both internal fuel and weapons carriage.[3] It represents the next-generation air-superiority fighter. The program officially entered E&MD in 1991 after a successful Milestone II DAB, which resulted in an Office of the Secretary of Defense (OSD) Acquisition Decision Memorandum (ADM) on August 1, 1991. The E&MD phase is currently stretched over 11 years (completion in 2002) because of funding limitations. Nine test vehicles are planned during E&MD; flight of the first one is scheduled for May 1997.

Lockheed Martin Aeronautical Systems (LMAS) and Pratt & Whitney (P&W) are the two prime contractors reporting to the Air Force through a System Program Manager (SPM), also referred to as the System Program Director (SPD), who is located at Wright-Patterson AFB, Ohio. Key team members are Lockheed Martin Tactical Aircraft

---

[2]U.S. Navy, "PMA-265 F/A-18E/F DAB Briefing (MSII [Milestone II])," Arlington, Va., May 6, 1992.

[3]USAF, "F-22 SPO Briefing to RAND," Wright-Patterson Air Force Base (WPAFB), Ohio, July 25, 1995; and LMAS, "LMAS F-22 Briefing to RAND," Marietta, Ga., August 8, 1995.

Systems (LMATS) in Fort Worth, Texas, and Boeing Military Airplanes (BMA) in Seattle, Washington. They operate under a commercial teaming agreement; a single team program office is located at LMAS, Marietta, Georgia. Both primes are performing on Air Force cost-plus-fixed-fee/award-fee contracts with values of approximately $11 billion (LMAS) and $1.9 billion (P&W).

## ARMY RAH-66

The RAH-66 Comanche will be the Army's new armed reconnaissance helicopter for the twenty-first century. Replacing the OH-58, OH-6, and AH-1 helicopters for the primary missions of armed reconnaissance and light attack, this twin-engine, lightweight, advanced-technology helicopter incorporates fly-by-wire flight controls, low-radar-signature design, a composite fuselage, and advanced mission equipment. Advanced mission equipment will include second-generation target-acquisition and night-vision sensors in an advanced electronics architecture. The program officially entered the demonstration/validation (Dem/Val) prototype phase with a successful Milestone I DAB in June 1988. Affordability considerations, budget reductions, and Service-/DoD-directed senior-DoD scope-of-work streamlining efforts have caused the program to be restructured and approved in its current scope: to build two prototype aircraft to undergo flight testing. The program also includes six early operational capability (EOC) aircraft that will be evaluated in the field before the Army seeks approval for initiation of E&MD and any follow-on low-rate initial production (LRIP). First flight of a prototype was scheduled for the end of November 1995 (it actually flew in early January 1996).

The airframe for the Comanche is being designed, developed, and built by a joint venture of Boeing Helicopters (BH) and Sikorsky Aircraft (SA), which have established an integrated program management team, called a Joint Program Office (JPO), to lead the contractor efforts at both BH and SA. This JPO reports to an Army Program Manager located in St. Louis, Missouri, at the Aviation and Troop Command (ATCOM). The engine is being developed and produced by the Light Helicopter Turbine Engine Company (LHTEC), a partnership of Allison (General Motors [GM]) and Garrett (Allied Signal). Boeing Sikorsky are performing under a cost-plus-incentive/award-fee contract, and LHTEC is under a cost-plus-incentive-fee contract. Contract values are approximately $2 billion for the airframe and $200 million for the engine.[4]

---

[4]U.S. Army, "RAH-66 PMO Briefing to RAND," St. Louis, Mo., August 16, 1995; Boeing Sikorsky, "Boeing Sikorsky JPO Briefing to RAND," Trumbull, Conn., August 29, 1995.

Chapter Four
# PROGRAM MANAGEMENT BY THE SERVICES

In this chapter, we present our research on and analysis of various aspects of the overall management of each of the three programs. We have tailored our discussions to the topics—criteria—identified in Chapter Two. The first section of this chapter, "Composite Program Management," summarizes our overall assessment. Following sections provide those details on each program that helped lead us to our overall assessment. Finally, the Appendix discusses a number of related processes that are unique to a particular Service or that are unique in how they are implemented, or are a "one-of-a-kind" approach, or documents something of interest to the authors. While not directly associated with our assessment, some processes are noteworthy because of their uniqueness or because of the amount of effort being expended by the program, or else deserve special treatment because they are being done in such depth, or require some detail to explain, or are believed important enough to this overall research to be documented in the detail provided in the Appendix.

## COMPOSITE PROGRAM MANAGEMENT

We consider all three programs to be well managed. Personnel, both military and civilian, on the government side are experienced, and individuals on industry's side are dedicated and motivated beyond profit to achieve program goals. Lessons learned from past programs are being taken seriously, and there is sharing of information on program-execution processes among the three government Program Offices. Openness and continuous communication among the contractor and subcontractor teams appear to be the norm now, as they do between the contractor teams and the government. Determination to avoid the mistakes of previous programs now appears to be a strong factor in this "no secrets" environment.

Rather than assess each of the three programs separately, which could create an unnecessary and misleading judgment regarding relative merit, proving an injustice to each Service, we have chosen to create an overall assessment (Table 4.1) for a composite view of all three programs. There is no one-and-only best way to manage. Rather, a combination of attributes, processes, and procedures in effect commonly within a Service and collectively within DoD can be used to create an environment and opportunity for good program management. The overall assessment was derived by the authors, using their experience and judgment about the organizations, management approaches, processes, and procedures in the Services and DoD today, and their qualitative belief in those changes that have occurred over the past 5–6 years, and whether those changes are for the good (or not) of fostering better program management. The ten criteria are not mutually exclusive; they intersect with and overlap the others to some extent. However, the matrix focuses on ten specific characteristics that should contribute to having a more successful program outcome than if any are excluded.

The rationale for these individual assessments is briefly stated in the table and is described in the various sections of this chapter. For example, for the "DPRO support has been instituted and is firmly established" criterion, we describe how we believe the use of the DPROs, their participation in all aspects of the program, and the use of program integrators (PIs) at the various DPROs as part of the PM's team have helped both to "link" (i.e., improve) DPRO communication with the PM and to exchange program data. We assessed this current practice as "good" and "improving," because the various Service PMs continue to look for ways to use the DPROs/PIs to assist them.

The cultural changes (e.g., acceptance by higher levels of management of bad news without fear of the messenger being killed or immediate retaliation against the program) over the past 3–4 years, in particular in DoD, will continue to improve the management of acquisition programs, unless external factors cause a regression to such

### Table 4.1
### Composite Management Assessment

| Criteria | Assessment | Rationale |
|---|---|---|
| Clear lines of authority have been established | Good to good-plus (Improving) | — DMR organizational relationship (SAE-PEO-PM) is apparent<br>— IPTs are used |
| Communication is encouraged | Good (Improving) | — Data are used and exchanged<br>— Various structured forums (written/verbal) |
| CS2, CPM, DAES, etc., are used | Good to good-plus (Improving) | — Used by management (government/contractor)<br>— Most IPTs are responsible for using these tools and do use them (cultural change)<br>— Real-time data are being used |
| Risk-management program/process is used | Fair-plus to good (Improving) | — Risk-mitigation techniques are actively used<br>— Different styles and emphasis at top levels [neutral] |
| Requirements are controlled | Good (Improving) | — DMR/PEO/PM have organizational control of requirements implementation[a]<br>— Requirements are stable<br>— Process for change is strict |
| DPRO support has been instituted and is firmly established | Good (Improving) | — DPRO is actively involved<br>— Participates at all levels<br>— Lead PI at DPROs for programs |
| Incentives are positive and apparent | Fair (Declining) | — Programs' being top priority of Services is motivation for success<br>— Schedules are budget-driven [disincentive] |
| Funding is stable; control and support are ensured | Fair-minus (Declining) | — Budgets have major instabilities<br>— Programs and/or contracts have experienced rephasings<br>— Support is lacking |
| Management team is selected for credibility and stability | Good (Improving) | — PMs have good backgrounds and experience<br>— DAWIA criteria are employed in selection<br>— PM-selection process is being formalized |
| Security promotes management involvement | Good (Improving) | — Security has no negative effects<br>— Controls are being reduced where possible |

NOTES: DAES = Defense Acquisition Executive Summary; DAWIA = Defense Acquisition Workforce Improvement Act.

[a]Users are not free to dictate changes at will. Block upgrades are made instead.

past practices as lack of openness, compartmentalizing of data, and diminished use of product teams (by both government and contractors).

Our main concern is with the unstable funding of these three high-priority programs. Without stable funding, DoD cannot demonstrate that a program is being effectively managed or is achieving the standard of excellence expected. All three programs tend to be the "Service reserves" for needed funds, or "cash cows," and intermediate staffs at all levels in government and Congress disregard the consequences of funding instability.

## CLEAR LINES OF AUTHORITY HAVE BEEN ESTABLISHED

All three programs are managed in accordance with DoD Directives, DMR Guidelines, and Service Implementation Regulations, Directives, and Instructions. All have a common reporting relationship, as depicted in Figure 4.1.

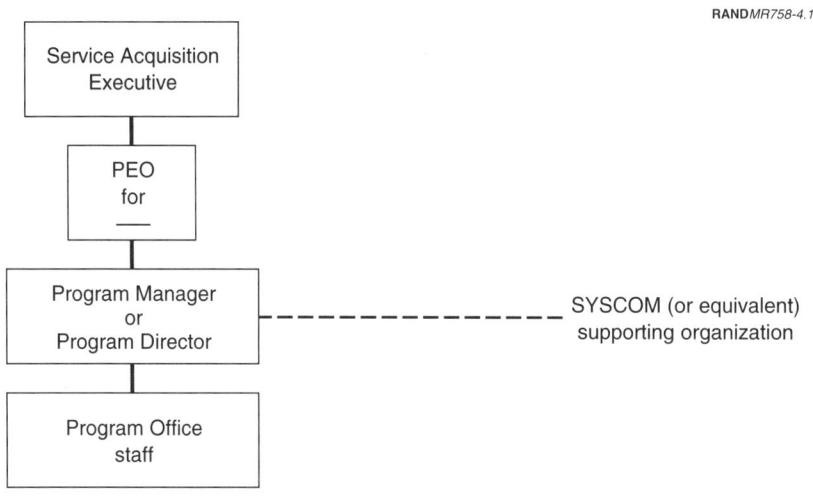

Figure 4.1—Service Management Model

## Navy F/A-18E/F

The Program Manager of the F/A-18, a Navy captain, reports directly to the PEO for Tactical Aircraft Programs (PEO [T]), who, in turn, reports directly to the Assistant Secretary of the Navy for Research, Development and Acquisition (ASN [RDA]). Formal written documentation describes the responsibilities and reporting chain for each of the three levels of management: SECNAVINST 5400.15 of August 5, 1991,[1] describes the responsibilities of the ASN (RDA); a formal charter[2] signed by the ASN (RDA) defines the responsibilities of the PEO (T); and a NAVAIR Instruction of 1982[3] defines the PM's (PMA-265's) responsibilities. Although the reporting chain of the NAVAIR document is outdated (due to the DMR SAE/PEO/PM reorganizations), the scope of PM responsibilities it defines still holds. (This document is currently being updated.) The Program Manager relies on NAVAIR to provide significant support from government headquarters and field activity, primarily through the Naval Air Warfare Center. Management is driven by the concept of a program and/or functional support team, or matrix-type activity. The use of integrated product teams (IPTs) is prevalent throughout the government and industry organizations. Formal NAVAIR government support is defined in an operating agreement signed by both the Commander, NAVAIRSYSCOM (COMNAVAIRSYSCOM), and the PEO(T), and is approved by the ASN (RDA).[4]

## Air Force F-22

The System Program Director of the F-22, an Air Force major general located at Wright-Patterson AFB, Ohio, reports directly to the PEO for Tactical and Airlift Programs in the Pentagon, who, in turn, reports

---

[1] USN, "DON Research, Development and Acquisition Responsibilities," Arlington, Va.: SECNAVINST 5400.15, August 5, 1991.

[2] USN, "Charter for the Program Executive Officer Tactical Aircraft Programs," Arlington, Va., August 16, 1990, signed by ASN (RDA).

[3] USN, "F/A-18 Program Charter," Arlington, Va.: NAVAIR Instruction 5400.74B, January 11, 1982 (currently being updated).

[4] Operating agreement between COMNAVAIRSYSCOM and Naval Aviation PEOs, Arlington, Va., August 16, 1990, approved by ASN (RDA).

directly to the Assistant Secretary of the Air Force (Acquisition).[5,6] Guiding documentation is the "F-22 Program Management Directive"[7] of April 18, 1994 (latest approved version; in process of annual update). The F-22 SPD is the total life-cycle manager of the F-22 Program. The SPD is called the Integrated Weapon System Manager (IWSM), because this person has been given total program responsibility and authority over logistics and test, as well as over those organizations supporting the program through the Sacramento Air Logistics Center's F-22 System Support Manager and the Air Force Flight Test Center's F-22 Combined Test Force.

Actual program execution by the SPD occurs through the concept of integrated product development (IPD), which was instituted by the Air Force Material Command. IPD is a "philosophy that systematically employs a teaming of functional disciplines to integrate and concurrently apply all necessary processes to produce an effective and efficient product that satisfies customer needs."[8] This teaming utilizes IPTs for day-to-day management within the F-22 System Program Office (SPO), also located at Wright-Patterson AFB. These teams are *product*-focused, which means that they are responsible for the performance, schedule, and cost (including risk management) of their products—what the SPD refers to as the *Iron Triangle* responsibility of the IPT. IPTs are formed at various levels of the program breakdown structure, which is referred to in terms of "tiers," Tier 1 being the total-weapon-system level and Tier 5 being the component/subsystem level. Because of the importance of this management approach (the use of IPTs for management, execution, and control, and the fact that the F-22 SPO is considered to be the leader in the use of IPTs in the Air Force), IPTs are discussed in more detail in the Appendix.

---

[5]Secretary of the Air Force, "Functions of the Secretary, Under Secretary, and the Assistant Secretaries of the Air Force," Washington, D.C.: Order 100.1, May 1990.

[6]USAF, "Acquisition System," AFPD 63-1, August 31, 1993.

[7]USAF, "F-22 Program Management Directive," April 18, 1994, signed by SAF/AQ.

[8]USAF, "Air Force Material Command Guide," May 25, 1993.

## Army RAH-66

The PM for the Comanche, an Army brigadier general located at ATCOM in St. Louis, Missouri, reports directly to the PEO for Aviation, also located at ATCOM, who, in turn, reports directly to the Assistant Secretary of the Army (Research, Development and Acquisition). Guiding documentation is Army Regulation 70-1.[9] The PEO for Aviation is operating under an appointment signed by the ASA (RDA);[10] the PM is operating under a charter signed by the PEO for Aviation.[11]

The RAH-66 is in Dem/Val, but because the PM has life-cycle responsibility, the general is also in charge of planning and execution of logistics and training (includes training systems), as well as other support activities. Within the PMO, the PM has a program staff of approximately 91 military and civilian personnel who are organized along functional lines. In addition, the PM receives program support (both reimbursable [PM pays] and nonreimbursable [host organization pays]) from ATCOM, the major supporting organization for the RAH-66 Program. Exact requirements are determined and updated in annual business plans (discussed in the Appendix).

## COMMUNICATION IS ENCOURAGED

All three programs have established structured (specifying timing, type of information, to whom) approaches to communicate verbally and in writing, and to report status at all levels of management, including between subcontractors and major primes/major team contractors, between major primes/major team contractors and government program offices, and between program offices and their higher-level leadership, the PEOs and the Service Acquisition Executives. Structured approaches also provide a means of sharing information with other involved Service organizations, i.e., the users of the products under development. Tables 4.2 through 4.4 briefly describe the type of DoD or Service reporting requirements with

---

[9]USA, "Army Acquisition Policy," Army Regulation 70-1, April 20, 1993.

[10]USA, "Appointment to the Position of Acting Program Executive Officer, Aviation," July 20, 1995.

[11]USA, "Charter of the PM, Comanche," September 27, 1994.

which the programs are complying, as well as some of the requirements the programs have instituted themselves. These tables are *not* all-inclusive but represent the type of reporting and communicating that is being accomplished, and at what levels, and indicate that such reporting and communicating are being done more openly (no secrets—all information is divulged, by different media and avenues) and more frequently than in the past.

We have purposely omitted some specifically mandated reports, such as the *Selected Acquisition Report* (SAR), because our focus was on internal DoD/Service procedures that are indicative of changes, openness, and a willingness and/or desire to bring everyone (government and contractor) into the communication loop. In addition, while each Service has issued a supplement to the DoD 5000-series directives and instructions, that supplement is intended to cover only necessary Service-specific aspects, not to dictate how to implement the DoD series. The constant use of video teleconferencing centers (VTCs), on-line management information systems, electronic networks, and daily telephone exchanges ensures that open communication (no secrets, real-time) for exchanging current information prevails.

## Navy F/A-18E/F

A significant set of requirements has been established to provide structured reporting of data, information, and program status between the contractors and the Navy, and also among organizational levels within the Navy. Table 4.2 lists the type of general reporting requirements with which the program is complying, as well as those requirements that are unique to the F/A-18E/F Program.

## Air Force F-22

The Air Force has set up a structured approach for communicating and reporting status at all levels of management and program execution: between subcontractors and major team contractors, between major team contractors, between major primes and the SPO, and between the SPO and higher-level leadership—the PEO and SAF/AQ—as well as other Air Force/DoD officials. Table 4.3 briefly lists and describes some of the reporting requirements that have been estab-

## Table 4.2

### F/A-18E/F Reporting Requirements and Related Activities

- DODD 5000.1, DODI 5000.2 and 5000.2M of February 23, 1991
  — Part 11—Program Control and Review
  — Part 16—DAES (Quarterly)
  — Part 19—Program Deviation Report
- NAVY SECNAVINST 5000.2A of December 9, 1992, implementing DOD 5000 series
- ASN (RDA) Memo of September 22, 1992, "Cost Performance Analysis Revitalization"
  — Includes identification of "Early Warning System" threshold criteria
- ASN (RDA) memos of October 24, 1994, and November 1, 1994, establishing process for ACAT I Program Reviews
- Weekly
  — Formal structured VTC meetings
    - between Northrop Grumman and MDA
    - between MDA and PMA-265
  — Earned-Value Analysis
  — PMA-265 "e-mail" status reports to key government/contractor personnel
- Monthly
  — PMA-265 status report (executive level) with TPM data to top Navy, fleet, contractor personnel
    - Risk Assessment Reports
    - Program Independent Analysis Briefings

lished for the program. To ensure that communication and discussions occur and are not set aside because of other perceived higher-priority considerations, specific times are set for many of the items that are on monthly or more-frequent schedules. Frequent use of VTCs, the on-line management/technical information system (M/TIS), and daily telephone exchanges ensures that openness prevails for exchanging current information. In addition, all-hands meetings are regularly scheduled as a means of keeping government and contractor employees informed at all levels of the SPO and contractor organizations.

## Army RAH-66

The RAH-66 PM has established a structured approach to facilitating communication, and to obtaining and passing on timely program information. This approach includes obtaining necessary information from the major subcontractors, the Boeing Sikorsky JPO and

## Table 4.3

### F-22 Reporting Requirements and Related Activities

- DODD 5000.1, DODI 5000.2 and 5000.2M of February 23, 1991
  — Part 11—Program Control and Review
  — Part 16—DAES (Quarterly)
  — Part 19—Program Deviation Report
- AF Supplement 1 to DODI 5000.2 and 5000.2M, August 31, 1993
- Weekly
  — SPO picture/telephone meeting with PEO/PEM
- Monthly
  — SPO Review of DPRO Program Status Data and Monthly Assessment Reports
  — Written Acquisition Report from SPO to SAF/AQ and PEO
  — CPR from primes to SPO
  — Supplier telecons (Tier 1 to supplier PMs/VPs)
- Bi-monthly
  — SPO meetings with contractor team (Tier 2) and company/sector presidents
- Quarterly
  — Formal program review between SPO and contractor team
- Semi-annual
  — PO meeting with contractor team CEOs
  — Supplier conferences (Tier 1 to supplier PMs/VPs)

prime contractor team, the engine prime contractor, the TSM, and the DPROs. Formal and informal reporting from the PM to his higher levels of management occurs frequently—daily with the PEO. Table 4.4 lists and briefly describes the key reporting requirements that have been established by both higher-level organizations (Army and Boeing Sikorsky) and the PM and JPO director.[12,13]

## CPM, C/SCS, DAES, ETC., ARE USED AT SERVICE ACQUISITION EXECUTIVE LEVELS

All three Service Acquisition Executives have emphasized the use of contract performance measurement (CPM) to their PMs. Use of the

---

[12]USA, "RAH-66 PMO Briefing to RAND," August 16, 1995.

[13]USA, "RAH-66 Boeing Sikorsky JPO Briefing to RAND," Trumbull, Conn., August 29, 1995.

### Table 4.4

### RAH-66 Reporting Requirements and Related Activities

- DODD 5000.1, DODI 5000.2 and 5000.2M of February 23, 1991
    - Part 11—Program Control and Review
    - Part 16—DAES (Quarterly)
    - Part 19—Program Deviation Report
- AR 70-1, "Army Acquisition Policy," March 31, 1993, and DA Pamphlet 70-3, "Army Acquisition Procedures," implementing the DOD 5000 series
- Daily
    - PM discussions with PEO
    - PM-to-PM telephone exchanges
    - JPO program integration team conference calls
- Weekly
    - First-flight VTCs and Significant Activity Reports (forwarded to Military Deputy to ASA [RDA])
    - JPO/PDT meeting
    - BH/SA presidents' Software Review with JPO
- Monthly
    - First Team (prime/key subcontractors) review technical, schedule, cost status with JPO
    - PMO/contractor CPR (cost performance report) and financial reviews
    - Army Acquisition Program Evaluation and Review System (AAPERS) status report to Army leadership
    - DPRO Assessment Reports to the PM
    - TSM forwards reports to the PM
- Quarterly
    - DAES report to PEO, ASA (RDA), and USD/A&T
    - "Face-to-face" between JPO and BH/SA presidents
- Semi-annual
    - Formal program reviews and Executive Steering Group meeting (government/contractor)
    - First team presidents (key subs) with BH/SA leadership and JPO

Cost/Schedule Control System (C/SCS; commonly referred to as CS2, which will be used throughout this report) is formally documented in directives or policy statements from the SAEs to their acquisition organizations. All three SAEs review the status of programs periodically, and participate in Service reviews of Defense Acquisition Executive Summary (DAES) reports prior to USD/A&T meetings. Table 4.5 summarizes key aspects of CPM and DAES activities in each of the Services.

Table 4.5

Use of CPM/DAES

| Party | Service | | |
|---|---|---|---|
| | Navy | Air Force | Army |
| Organization responsible for CPM/DAES at Service level | ASN (RDA) | SAF/AQ | ASA (RDA) |
| SAE actively involved with CPM/DAES? | Yes | Yes | Yes |
| SAE staff office identified for providing and monitoring assistance to SAE? | Yes (OASN [RDA] for Resources and Evaluation) | Yes (SAF/AQX and SAF/AQP) | Yes (OASA [RDA] for Program Evaluation) |
| PMs and/or PEOs participate in SAE CPM/DAES reviews? | Yes | Yes | Yes |

Each of the three programs manages differently, as is to be expected, and each emphasizes different techniques to control its programs. Common threads run through the three, however, including

- completely open communication (no secrets—all information divulged, through a variety of media and forms) between the government and contractor teams.

- use of IPTs/PDTs (product development teams) to manage product-focused areas, including technical, schedule, and, for the F/A-18E/F and F-22 Programs, responsibility of these teams for managing their allocated portion of budgeted costs.

- sharing of formal and informal data, as near real-time as possible, with their available information systems and electronic networks.

- bringing major subcontractors into top management teams, and obtaining and using CS2 data from these subcontractors in near real-time.

- major emphasis on use and tracking of technical performance measures (TPMs).
- use of reports, not just pro forma but as active tools to help government-contractor management teams track program progress.

## Navy F/A-18E/F

Within the Navy, the formal use of CS2 is documented in SECNAVINST 5000.2A of December 9, 1992,[14] which establishes the requirements and reporting procedures, and names the Office ASN (RDA) for Resources and Evaluation as the focal point. Active use of the CPM and the DAES report by the ASN (RDA) is documented in a September 1992 memo[15] that establishes reporting requirements and use of these data, as well as establishing an early-warning system for out-of-cycle reporting of threshold breaches. (*Out-of-cycle reporting* refers to the PM's obtaining program information, between DAES reports, of a performance, schedule, or cost estimate that exceeds limits established in the program baseline document and/or limits established in the DOD 5000–series documents.) ASN (RDA) memos of October 24, 1994, and November 1, 1994,[16] reinforced the ASN (RDA)'s desire to conduct major program (ACAT 1) reviews, with an emphasis on the CPM. Formal schedules have been set for all programs to submit DAES reports to the Navy, and on to OSD, on a quarterly cycle. Monthly reviews are held within the Navy secretariat; reviews are held with the OUSD/A&T on reports scheduled for that particular month.

One of the key tools used by the Navy Program Manager to manage this program is the on-line near-real-time management information system of the prime airframe contractor, MDA. Referred to as IMICS (integrated management information and control system), this sys-

---

[14]USN, "Navy Implementation of DoD 5000 Series," Arlington, Va.: SECNAVINST 5000.2A, December 9, 1992.

[15]USN, "Cost Performance Analysis Revitalization," Arlington, Va.: ASN (RDA) Memo, September 22, 1994.

[16]USN, "Contractor Performance Management Reviews," Arlington, Va.: ASN (RDA) Memo, October 24, 1994; USN, "Program Reviews," Arlington, Va.: ASN (RDA) Memo, November 1, 1994.

tem tracks some 6,000 sets of data, is updated regularly (sometimes weekly), and is shared real-time between MDA and the government. This open-communication linkage facilitates interactive management of the program. One set of "books" is being used by all involved parties.

Formal use of CS2/CPM is another key tool. The contractor has proven it to be an important management tool down to the fourth and fifth tiers of the program Work Breakdown Structure (WBS). Engineers responsible and accountable for cost and schedule at these levels regularly use, monitor, and report (weekly) on earned-value/CS2 status. In the opinion of both the Navy Program Manager and the MDA Program Manager, this attention to cost detail has resulted in significantly greater attention to cost control at the level necessary to be successful.

Within CPM, hard data on key technical performance parameters, cost, and schedule are routinely reported on, as are other important program information, including a summary of program independent analysis (PIA) topics (discussed in the Appendix). Briefly, in a PIA, a small team or teams of individuals in the Navy PM's office and the contractor PM's office conduct short (timewise) special reviews and investigations of critical program topics identified by either the Navy or contractor PMs. Because it verifies whether what is being done is correct, a PIA is believed to be important for successful program completion (performance, schedule, and cost). Continuous tracking of technical performance over time is widespread and can be used to alert management to unfavorable status or trends of these parameters. Figure 4.2 shows one example, the status of empty weight over a 16-month period.[17] Empty weight is a critical parameter for an aircraft program. Historically, in most programs, empty-weight estimates increase during the development phase of acquisition, decreasing mission performance (range and/or payload). Empty-weight status is a key parameter tracked closely by both government and industry leaders, although each program may have a different process for tracking it.

---

[17]PMA-265 (USN), "FA-18E/F Program Update Briefing" (to RAND), Arlington, Va., January 26, 1995.

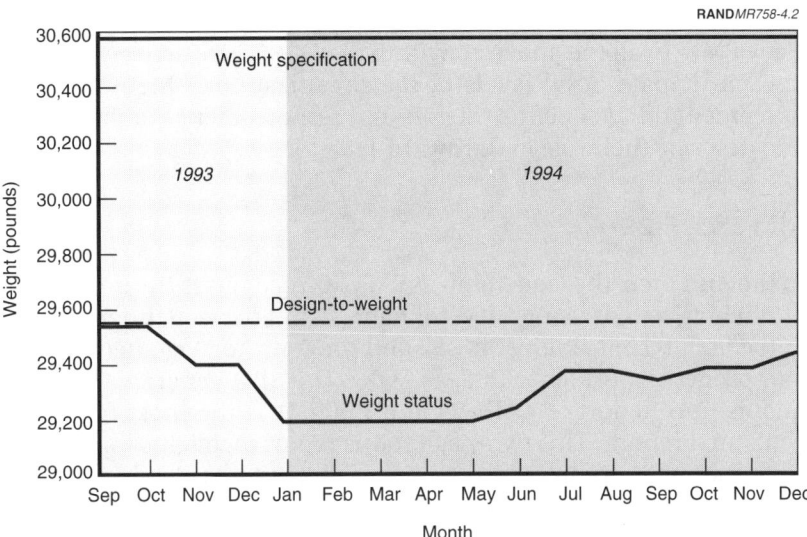

SOURCE: PMA-265 (USN), "F/A-18E/F Program Update Briefing" (to RAND), Arlington, Va., January 26, 1995.

Figure 4.2—Empty-Weight Technical Performance Measure for the Navy's F/A-18E/F Program

An award-fee provision is included in the contract to emphasize particularly important portions of the program, key events, or elements of the Statement of Work, and/or to provide an incentive for the contractor team to focus adequate attention on those items the government PM considers most important or requiring special attention.

The Program Manager considers periodic award-fee evaluations of both airframe and engine contractors to be effective feedback (communication) to the contractor on how the program is proceeding: This information, especially since profit (award fee) is always reported up the corporate management chain, becomes an added incentive for the contractor to do well in the areas highlighted in the award-fee plan. Such feedback, along with the dollars associated with the award-fee performance periods, ensures a formal loop (written communication) and understanding between government and industry. Members of the Navy Support Team to the Program provide written input to the Award-Fee-Determining Official (the Navy PM). In all the above activities, the DPROs are actively involved and participate at all levels of the organizations. Memoranda of Agreement (MOAs) define the role and responsibilities of the DPROs, and describe their role in the award-fee process.

## Air Force F-22

In the Air Force, the leadership at SAF/AQ recognizes the importance of tracking and of using CPM to manage and oversee programs. Air Force Staff responsibility for CS2 and program-management reviews and processes resides with SAF/AQX. The SPD-generated Monthly Acquisition Report to SAF/AQ and PEO/TA is the most frequently generated report. This two-page status report highlights top-level issues on the program to the Air Force acquisition leadership. Formal schedules for the quarterly cycle Air Force reviews of DAES have been established by SAF/AQ, as has the process to be followed for review of current data. The PEM provides this information to SAF/AQ during what is called the Acquisition Program Review Board meetings, at which the PEO/TA or a staff member is present. Semi-annually, the PEO/TA portfolio review is held with SAF/AQ, and all the programs under the PEO are reviewed at one time with SAF/AQ. The acting SAF/AQ Memo of May 12, 1995, "Use of Earned Value on USAF Programs," stresses the importance of CS2 and earned value, and how earned value will be presented at the PEO portfolio review. The SPD participates by providing current program status and issues to the Air Force Service Acquisition Executive.

The F-22 SPD utilizes a number of techniques to control the program; no single technique predominates. IPTs and their feedback mechanisms are the foundation for controlling the F-22 Program. The various IPTs have product responsibility and authority over technical performance, schedule, and cost (allocated budget). Openness, and shared communication and data are key. Each IPT tracks progress with common tools, which include the Requirements Traceability Management (RTM) report; a system maturity matrix (SMM); appropriate technical performance measures (TPMs); the integrated master plan (IMP), which is traceable to the integrated master schedule (IMS); and appropriate CS2 data. (These items are a set of documented Air Force common processes that are utilized throughout Air Force program offices. Brief descriptions are given in the Appendix.) These items provide feedback to the immediate IPT leader and the next-higher-tier IPT leader, on executing the IPT/WBS SOW task against the requirements of the specification paragraph, the integrated master plan and integrated master schedule requirements, and the allocated budget for that activity. Such tracking metrics at the various IPT tier levels are aggregated to obtain the Tier 1 weapon-system-level status. Approximately 236 TPMs are available on-line through the M/TIS, are tracked over time, and are used to provide valuable information to both the SPO and contractor management. Figure 4.3 gives an example of how one TPM, empty weight versus time, is tracked for this program.

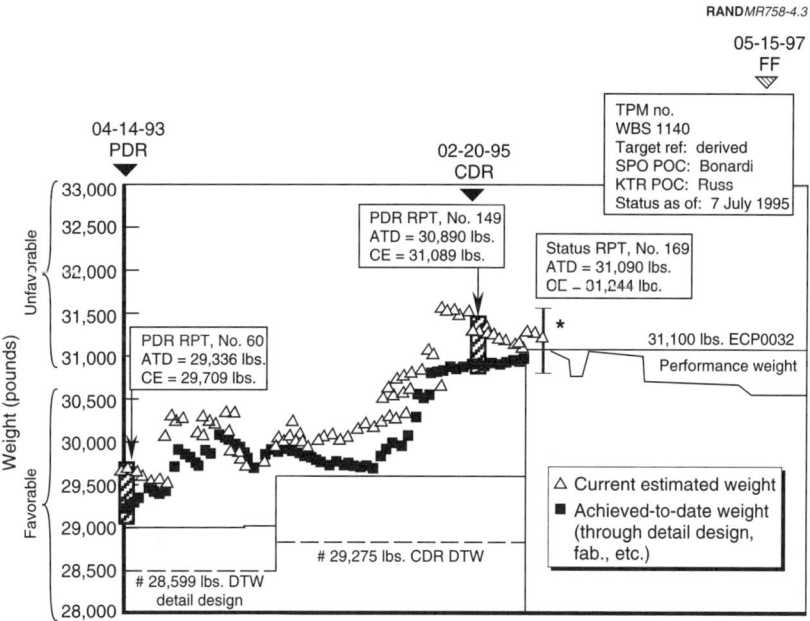

SOURCE: USAF, "F-22 SPO Briefing to RAND," WPAFB, Ohio, July 25, 1995.

Figure 4.3—Empty-Weight (Less Engines) Technical Performance Measure for the Air Force F-22 Program

With regard to schedule, the program utilizes the integrated master plan and the integrated master schedule as controlling and tracking tools. The IMP is a contractor-prepared event-based plan of activities that must be accomplished and the criteria needed to determine activity success. Approximately 12,000–13,000 IMP activities and events on the air-vehicle contract are described and tracked. The IMP is a contractual item; the IMS is not. The IMS is a required Contract Data Requirements List (CDRL) item that tracks, on a time-based schedule, some 30,000 tasks based on the IMP. Thus, the program manages events and activities that are contractually binding but are not tied contractually to a schedule; for a variety of reasons, a binding schedule would require frequent contract changes (see "Funding Is Stable . . ." criterion below for a discussion of the effects of scheduling changes).

The SPD considers E&MD cost control and production design-to-cost (DTC), now termed production cost estimate (PCE), to be important to the future success of the F-22. Cost performance reports (CPRs) contain CS2 cost data for the previous month for major team members *and* major suppliers, rather than the usual one-month lag on subcontractor data—which gives the SPO the most current information possible. The SPO not only tracks and evaluates these two items (E&MD and DTC data) but also incorporates the data into a cost-estimating model, called the production cost model, to project learning-curve estimates using pre-production verification and initial production aircraft experience.

In addition to the open and frank feedback and communication between the Air Force SPD and prime air-vehicle and engine contractors, the award-fee provisions of the E&MD contract offer the primary motivation to the contractors and account for the major portion of their profit. The PEO/TA, the Award-Fee-Determining Official for both contracts, receives input monthly through Award-Fee Board meetings, in which Air Force IPTs provide data against the criteria for the current six-month period of performance. Mid-term assessments are provided to the contractors, and final written evaluations are provided at the time of the award-fee determinations. The DPROs are actively involved in all of these activities as part of the various IPTs at each tier.

To integrate all these data, the SPO is developing what it calls an F-22 Information Resources Management (FIRM) management/technical information system. When fully developed, procured, and deployed, it will link the LMAS M/TIS, the system/software engineering environment data, the video teleconferencing system, and an expanded data network among all major sites, including all SPO locations, Air Force users, prime contractors, and major subcontractors. Currently, on-line sharing of complete M/TIS data occurs between the SPO and the prime contractors, and TPMs, IMP/IMS data, CS2, and other metrics are shared in real-time.

## Army RAH-66

The Army relies heavily on analyzing technical, schedule, and cost data. Requirements for formal CS2 reporting are contained in various Army documents. DA Pamphlet 70-3, "Army Acquisition Procedures," describes the reporting requirements and names the Director for Program Evaluation as the point of contact (POC) in the Army Acquisition Executive's staff. Recent support for earned value has been promulgated by a memorandum[18] to PEOs and other acquisition organizations. Guidance on reporting program status data to the ASA (RDA) and immediate staff include the monthly Army Acquisition Program Evaluation and Review System (AAPERS) from the PM, the monthly acquisition program review (MAPR) process,[19] and the quarterly DAES report.

These data and reports are reviewed in a number of staff offices; consolidated findings are reviewed by the Director of Assessment and Evaluation, who, in turn, summarizes key issues to the ASA (RDA). In support of the ASA (RDA), CPR data are also reviewed independently by Army Materiel Command Headquarters, as well as through the DPRO/DCMC chain. The ASA (RDA) level establishes formal schedules for the quarterly cycle Army reviews of DAES and reviews documentation and data prior to the USD/A&T reviews.

---

[18]USA, "Earned Value Management," ASA (RDA) Memorandum, August 2, 1995.

[19]USA, "Monthly Acquisition Program Review (MAPR)," Military Deputy to ASA (RDA) Memorandum, August 3, 1993.

The Comanche PM relies on a number of processes and means for running the program; no one item or system can be pointed to as the central focus. Because this is a Dem/Val program, the PM integrates various separate sets of data to continually assess program status.[20] The PM relies on the PMO staff, various government development teams (GDTs), and subject-matter experts to interface with the Boeing Sikorsky JPO Product Development Teams (PDT), track program events and status, and review contractor data. Essential to team progress-tracking and accountability are use of CS2, requirements allocation (from the Operational Requirements Document to weapon-system specification, and, further, to the lowest-level component specifications), technical performance measures, and timely access and feedback of data. The PM holds monthly reviews with the PMO staff and JPO Director or Deputy to review and discuss these data.

The JPO utilizes a series of PDTs within BH and SA to manage their product areas for technical progress and schedule (allocated budget remains a functional responsibility). Interlinking airframe-and-armament and mission equipment package PDTs connect these component PDTs at each company location, and an integration PDT at the JPO level coordinates PDTs across company lines. A substantial set of TPMs (approximately 85) is being used and tracked. Figure 4.4 is an example of the TPM for empty weight.

CS2 is tracked on paper (instead of on-line, real-time), which is delivered through the monthly CDRL requirement for CPRs. Recognizing the delay of both prime and subcontractor data, the JPO uses other means to provide current data, including weekly scheduled meetings to review current status; updating available CS2 information, particularly from their "First Team" (this term refers to the name of their major subcontractors) of the top 15 subcontractors; and providing a three-month forecast. Manpower data are reviewed weekly by the JPO.

The key to a management information/control system, we were told, is having a reporting system that is useful to management. Both the

---

[20]USA, "RAH-66 PMO Briefing to RAND," St. Louis, Mo., August 16, 1995.

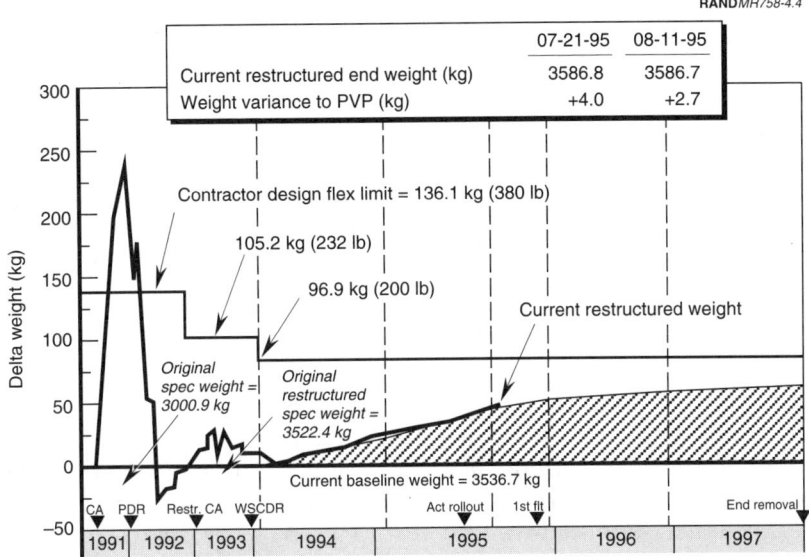

SOURCE: USA, "RAH-66 PMO Briefing to RAND," St. Louis, Mo., August 16, 1995.

Figure 4.4—Empty-Weight Technical Performance Measure for the Army RAH-66 Program

JPO and the Army PMO consider CS2 to be part of such a management tool. The PMO holds monthly meetings to discuss the performance, cost, and schedule status; independent assessments are also made by both ATCOM and Headquarters, Army Materiel Command subject-matter experts.

Like PMs for the Navy and the Air Force, the RAH-66 PM considers the use of an award fee to be a positive motivator for this program. Award-fee provisions of the airframe contract are another tool the PM uses to manage the program. Four percent of the contract target cost has been set aside as the potential award fee. The PEO is the official who determines the award fee, and the PM is the chairperson of the Award-Fee Board. The TRADOC Systems Manager and Program Integrator at the JPO DPRO are members of the board, as are other senior members of the PMO and ATCOM, which enables the PM to utilize the knowledge base from the DPRO staffs' full-time participation at the contractors' facilities.

The PM is in the process of developing an automated management information system to provide for real-time data exchange with the JPO and other government organizations. Internet and limited electronic transfer of data are available, primarily in the supportability and software areas. A flight-test data module is in the process of coming on-line to track test information, status, and corrective actions.

Within the Army structure, the PM relies on the Team Comanche concept (discussed in the Appendix) of using three levels of management. Within this structure, process-action teams, working with all involved organizations, both government and contractor, cover such areas as cost reduction and cost avoidance, first flight, performance, simulation, and testing. PATs are also formed to review any special areas of concern the PM may feel warrant an intense, short-duration investigation.

## A RISK-MANAGEMENT PROGRAM/PROCESS IS USED

Risk management is an important part of all three programs, although each Service handles it differently. In fact, the risk-management approaches are probably the most diverse of any of the aspects assessed. Each program defines *risk-management* differently and uses various methods to track mitigation efforts. Each approach is unique within its own Service and program culture, and each program's PM believes its respective method to be an effective way to manage risk.

### Navy F/A-18E/F

Risk management is a key part of this program and is actively endorsed by top government and contractor program-management officials, who use the risk-assessment results to help focus their attention on potential problems. Contractual requirements call for documented airframe and engine plans.[21] The program has developed a structured process managed by a risk-assessment board chartered by the Navy PM. The board meets quarterly, but reports monthly, in writing, on its activities. Membership consists of the Navy Program Office, NAVAIR matrix, DPROs, and the contractors.

The risk-management program is based on a traditional four-step approach of identification, analysis, planning and/or handling, and tracking. A risk assessment of *high, medium,* or *low* is based on five levels of uncertainty and five levels of consequences, and each risk item is assessed with this matrix. Figure 4.5 shows the assessment criteria. The figure includes definitions of *risk* and *risk management* for this program.

Figure 4.6 is a recent F/A-18E/F Program risk-assessment chart showing the top program risks.[22]

## Air Force F-22

The SPO does not use a separate approach to risk management, but believes each of the program-management processes to be part of the risk-mitigation effort. Collectively, the SPO considers various documents, technical performance measures, and other tracking procedures, as well as the production readiness review (PRR) process, to be a risk-management approach.

Risk management is considered an integral part of the F-22 E&MD Program and is an integrated part of each IPT's Iron Triangle responsibility. IPTs are charged with identifying their own risks as early as possible, determining the cause and significance, and developing and implementing effective risk-mitigation actions. Individual IPTs report on these actions to their next-higher-level IPT. Each IPT has procedures for resolving these risks itself or can refer to a higher

---

[21]USN, "PMA-265 F/A-18E/F Program Risk Assessment Board Charter," Arlington, Va., undated; USN, "MDA F/A-18E/F Risk Management Plan," Arlington, Va., July 1, 1992; "GE F414-GE-400 Risk Management Plan," March 8, 1993.

[22]USN, "PMA-265 F/A-18E/F Program Update Briefing to RAND," Arlington, Va., January 26, 1995.

40   Three Programs and Ten Criteria

RAND*MR758-4.5*

Definitions:
- *Risk management*: an organized, systematic decision-making process that efficiently identifies risks, assesses or analyzes risks, and effectively reduces/eliminates risks to achieving program goals.
- *Risk*: an undesirable situation or circumstance that has both a probability of occurring and a potential ill-consequence to program success.

| Level of Risk | | Your Approach/MDA and NAD Processes |
|---|---|---|
| 1 | Not likely | • Will mitigate this risk |
| 2 | Low likelihood | • Have *usually* mitigated this type of risk with minimal oversight in similar cases |
| 3 | Likely | • *May* mitigate this risk, but workarounds will be required |
| 4 | Highly likely | • *Cannot* mitigate this risk, but a different approach *might* |
| 5 | Near certainty | • Cannot mitigate this type of risk; *no* known processes or workarounds are available |

■ High/major disruption of the plan. Management unlikely to alter outcome.
▨ Moderate/some disruption in the plan. Effective management actions possible.
☐ Low/little or no disruption. Current approach sufficient.

Given that the risk is realized, what is the magnitude of the impact on—

| Level | Technical | Schedule | Cost |
|---|---|---|---|
| 1 | Minimal or no impact | Minimal or no impact | Minimal or no impact |
| 2 | Moderate reduction; same approach attained | Additional activities required. Able to meet need dates | Budget increase <5%, or united production budget impacted |
| 3 | Moderate reduction, but workarounds available | Minor slip in key milestones. Not able to meet need dates | Budget increase >5%, or other teams impacted, or can't meet unit production budget |
| 4 | Major reduction, but workarounds available | Key milestones slip >1 month, or program critical path impacted | Budget increase >10%, or E&MD budget >2% |
| 5 | Unacceptable; no alternatives exist | Can't achieve key team or major program milestone | Budget increase >15%, or E&MD budget >10% |

SOURCE: USN, "PMA-265 F/A-18E/F Program Update Briefing to RAND," Arlington, Va., January 26, 1995.

**Figure 4.5—F/A-18E/F Risk Management Identifies Five Levels of Uncertainty and Five Levels of Consequences**

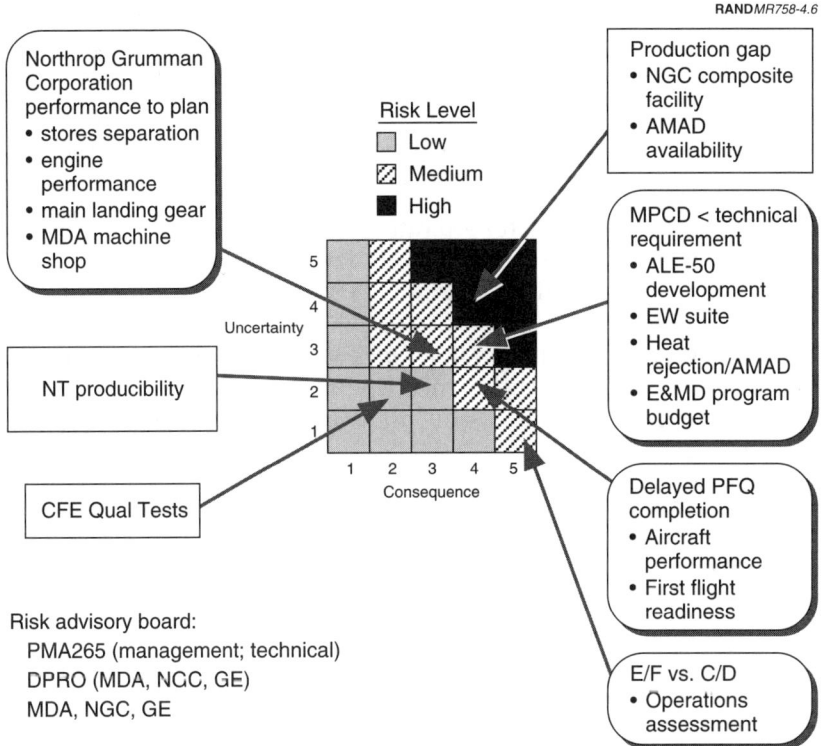

Figure 4.6—Risk Assessment of the E&MD Phase for the Navy F/A-18E/F Program

tier for help. Each week, the SPD updates the top-ten issues list at the Tier 1 level.

The team identifies the three basic causes of risk as lack of (1) understanding of the requirement, (2) mature technology to satisfy that requirement, and/or (3) a planning and tracking system to measure progress. The team response is a common set of plans, processes, and controls constituting what is called a *common language across the government/contractor team.*[23] This common language includes specifications and a requirements-traceability matrix, the Statement of Work, Work Breakdown Structure, integrated master plan, integrated master schedule, Cost/Schedule Control System, the system maturity matrix, technical performance measures, and the award-fee plan.

Of the 236 prime-contractor aircraft TPMs that are used and tracked, 15 are Tier 1 (weapon-system-level) TPMs and 53 are Tier 2 (air-vehicle-level) TPMs. In addition to TPMs, the IPTs track schedule and cost using various metrics. For example, the air-vehicle IPT is tracking IMS performance to first flight of the first prototype aircraft. Figure 4.7, which shows this metric,[24] indicates items started and items completed, and notes how many are delinquent (late in completion).

Following the policies of DODI 5000.2, the SPO discontinued the previous concept of large, separate teams doing the risk assessment over a concentrated 5–10-day period in favor of having the IPTs perform the assessment as an integral part of their critical design review activities. Instead, the SPO has implemented the concept of incremental production readiness reviews, in-process reviews that constitute a system- or subsystem-level risk assessment. The first PRR was conducted in conjunction with the CDR. This initiative, to do initial PRRs early in the development phase—to analyze potential production or producibility problems—is a significant cultural change over past programs. Changes in design are most easily made (and are less costly) early in a program rather than later.

The metric showing their initial risk assessment at the Tier 2 air-vehicle level is shown in Table 4.6. From this table, it can be seen that there are no high-risk items, but seven items have been evaluated as medium (*M*) risk.

---

[23]USAF, "F-22 SPO Briefing to RAND," WPAFB, Ohio, July 25, 1995.

[24]USAF, "F-22 SPO Briefing to RAND," WPAFB, Ohio, July 25, 1995.

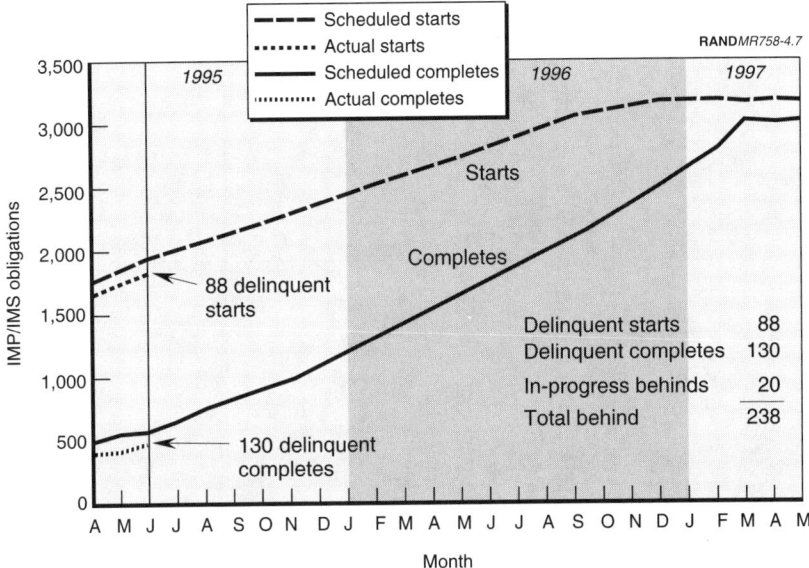

SOURCE: USAF, "F-22 SPO Briefing to RAND," Arlington, Va., July 25, 1995.

Figure 4.7—Air-Vehicle Integrated Master Schedule Performance to First Flight

## Army RAH-66

The PM utilizes several management tools, processes, and organizations to ensure that technical, schedule, and cost objectives are tracked and resolved. A formal risk-management process is in place. Based on the Defense Systems Management College (DSMC) model, it covers all weapon-system elements including the air vehicle, mission equipment package, propulsion, software, diagnostics and integration, supportability, producibility, and cost. The PMO, supporting government organizations, and the prime contractors all

## Table 4.6
### Initial Air-Vehicle PRR Assessment

| Product Design | Rate | Industrial Resources | Rate | Production Engineering and Planning | Rate | Materials and Purchase Parts | Rate | Quality Assurance | Rate | Logistics | Rate |
|---|---|---|---|---|---|---|---|---|---|---|---|
| Producibility assessment | M | Capacity | L | Manufacturing plan | L | Bill of materials | L | Structure and organization | L | Spares capacity | L |
| Design stabilized | M | Facilities and equipment | L | Production schedules | L | Make or buy | L | Quality-assurance program compliance | L | Support/test/diagnostic equipment | L |
| Design validated | L | | | | | | | | | | |
| Incomplete design risks | M | Manufacturing management systems | L | Methods/processes | L | Long lead | M | Procurement/quality acceptability criteria | L | Training equipment | L |
| System configuration | L | Personnel projections | L | Continuous improvement | L | Sole source | L | Participate in design and planning | L | Spare integrated with production | L |
| Operations/maintenance/support | M | Training and certification | L | Manufacturing instructions | L | GFM/GFE | L | | | | |

Table 4.6—continued

| Product Design | Rate | Industrial Resources | Rate | Production Engineering and Planning | Rate | Materials and Purchase Parts | Rate | Quality Assurance | Rate | Logistics | Rate |
|---|---|---|---|---|---|---|---|---|---|---|---|
| Technical data | M | | | Configuration management | L | Materiel control | L | | | | |
| Standardization | L | | | Production/ cost analysis | L | Procurement plan | L | | | | |
| Critical and scarce materials | L | | | Management information system | L | | | | | | |
| Foreign/ diminishing sources | L | | | Work measurement | L | | | | | | |
| Alternate sources | L | | | | | | | | | | |
| Production-cost projection | M | | | | | | | | | | |

SOURCE: USAF, "F-22 SPO Briefing to RAND," WPAFB, Ohio, July 25, 1995.

L—Low, M—Medium, H—High.

46  Three Programs and Ten Criteria

participate. The baseline for the risk-management approach was set in June 1991 with a formal signed risk-management plan.[25]

The PM's concept is to formally document risk in periodic reports tied to major milestones or program events, such as restructuring or streamlining efforts. Formally documented risk assessments were made in August 1992[26] and December 1993,[27] and in March 1995[28] following an update to the Risk Management Plan/Methodology in August 1994.[29] An updated Risk Assessment is planned for the second quarter of 1996. Within these plans and assessments are methodologies for assessing risks, including quantification on a 0-to-1.0 scale (with 1.0 being the highest risk) of both potential for failure and consequences of failure. Figure 4.8 illustrates the methodology used and the quantification numbers for high, significant, moderate, minor, and low risks.[30]

The technical staff in the PMO is responsible for managing the risk program for the PM. The key mechanism they use to track program status, problems, and issues is government development teams, whose membership includes PMO, Aviation Research, Development, and Engineering Center (AVRDEC), DPRO, and other Army and DoD organizations with subject-matter experts. These GDTs interface with the contractor PDTs and DPRO counterparts to track program progress against the risk-management plans that are in place. To review what has been accomplished, issues, and cost and schedule variances, internal PMO program reviews are held monthly. Similar topics and data are discussed at periodic contractor program reviews. PDTs with their counterpart GDTs follow and track the miti-

---

[25]USA, "RAH-66 Comanche Program Risk Management Plan," St. Louis, Mo., June 21, 1991.

[26]USA, "RAH-66 Comanche Program Risk Assessment," St. Louis, Mo., August 1992.

[27]USA, "RAH-66 Comanche Risk Management Plan," St. Louis, Mo., December 1, 1993.

[28]USA, "RAH-66 Comanche Risk Assessment Program (Draft)," St. Louis, Mo., March 30, 1995.

[29]USA, "RAH-66 Comanche Risk Management Program (Draft)," St. Louis, Mo., August 24, 1994.

[30]USA, "RAH-66 Comanche Risk Management Program (Draft)," St. Louis, Mo., August 24, 1994.

Program Management by the Services 47

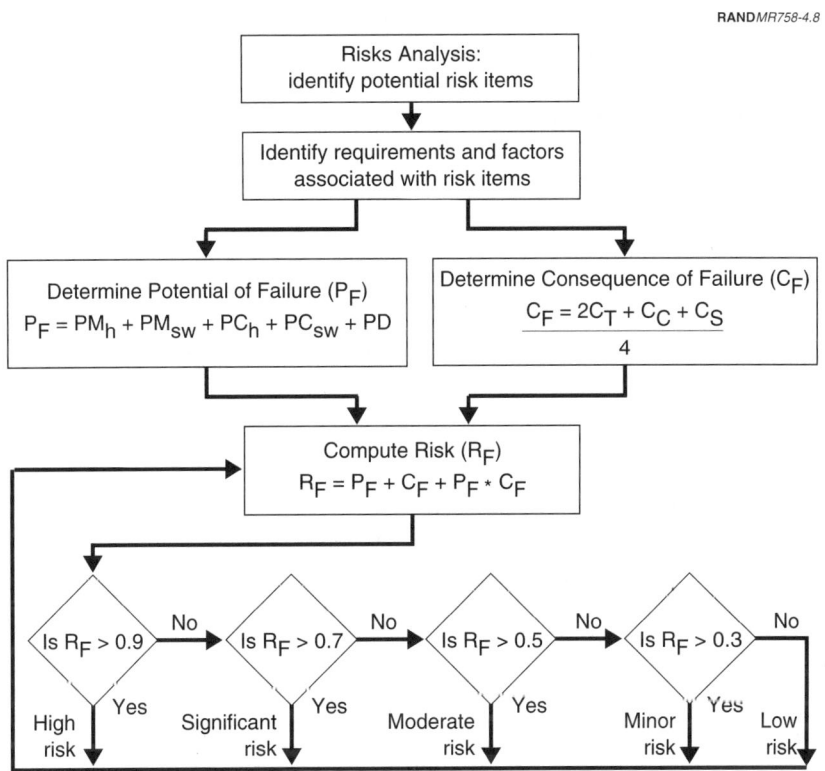

SOURCE: USA, "RAH-66 Comanche Risk Management Program (Draft)," August 24, 1994.

NOTES: $P_F$ = probability of failure
$R_F$ = risk factor
$C_F$ = consequence of failure
$C_T$ = consequence of failure due to technical factors
$C_C$ = consequence of failure due to changes in cost
$C_S$ = consequence of failure due to changes in schedule
$PM_h$ = probability of failure due to degree of hardware maturity
$PM_{sw}$ = probability of failure due to degree of software maturity
$PC_h$ = probability of failure due to degree of hardware complexity
$PC_{sw}$ = probability of failure due to degree of software complexity
$PD$ = probability of failure due to dependency on other items.

Figure 4.8—Risk-Assessment Flow for the Army RAH-66 Program

gation efforts established for each element according to the schedule for the particular element. Figure 4.9 shows the summary evaluation from the March 30, 1995, assessment (draft) report.[31] It shows that the mission equipment package (MEP) has the greatest risk of the various groups—.503—but is a moderate-risk item.

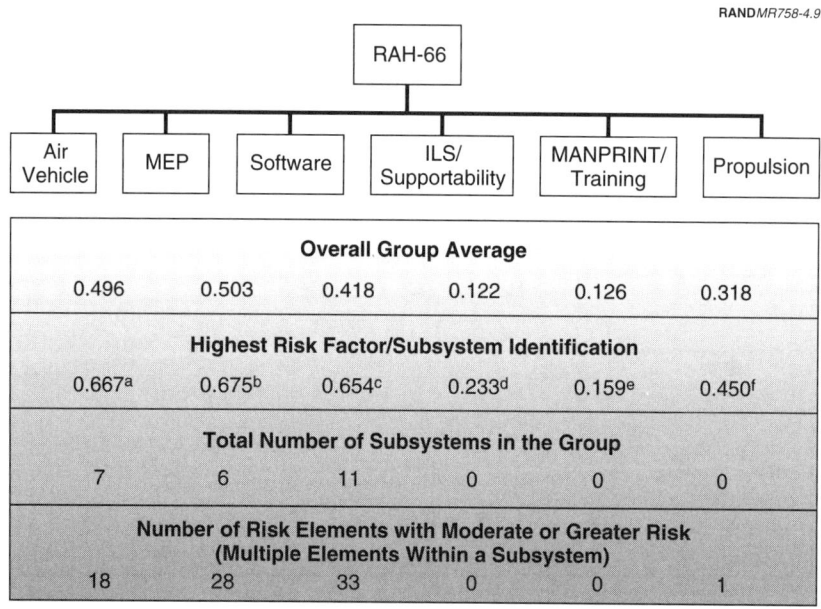

Figure 4.9—Risk-Assessment Summary for the Army RAH-66 Program

---

[31]USA, "RAH-66 Comanche Risk Assessment Program (Draft)," St. Louis, Mo., March 30, 1995, pp. 1–2.

## REQUIREMENTS ARE CONTROLLED

In many past acquisition programs, the making of numerous changes to operational (user) requirements after an acquisition program had begun has been criticized, both from within DoD and from outside individuals and organizations. As technology advanced and new improvements or capabilities came to the fore, users of weapon systems often said, "Yes, I want them on my xyz system," and the materiel developer (the acquisition program PM), who wanted to please his or her customer, usually said, "OK," then started worrying about cost and/or schedule effects. Over the past 5–6 years, attention within DoD has been focused on stabilizing requirements after the operational requirements have been identified and approved. Upgrades to a particular weapon system are considered only through a preplanned product-improvement program.

In addition to close cooperation among the developers and the users, and open (no secrets) communication, all three programs have stable requirements. The SAE-PEO-PM Defense Management Review relationships established, coupled with strong user representatives who understand the acquisition processes and implications, particularly in an austere budget environment, have assisted in this stabilization.

### Navy F/A-18E/F

It appears that the ASN (RDA)/Chief of Naval Operations (CNO) (N-8/N-88) relationships (interactions) are working well. The recent addition of a military three-star flag officer as Principal Deputy Assistant Secretary of the Navy (Research, Development and Acquisition) (PDASN [RDA]) will further facilitate current military-civilian interactions. The Navy will now have top acquisition leadership (civilian assistant secretary and military principal deputy) similar to the Army and Air Force.

Specific F/A-18E/F operational requirements appear to be stable, inasmuch as we were told that there had been only one change in user requirements (to add a new weapon to be integrated on the aircraft) since the DAB, and that that change followed a formal written process. The fleet is actively involved in the program, and key flag officers are regularly (i.e., weekly and monthly) informed of major activities and program status.

## Air Force F-22

The relationship between the user, represented by the Air Combat Command (ACC), and the Air Staff (AF/XOR) and the PEO/SPO is also working well. Reviews with corporate CEOs and Senior Air Force officials are held regularly. In addition, ACC is responsible for changes to the operational requirements, including obtaining funding of any changes that affect cost. The process followed for making changes in operational requirements is in full compliance with the DMR. We were told that, to date, only one major change has been made by the user since Milestone II; that change was the addition of the Joint Direct Air Munition (JDAM) as a weapon for the F-22.

## Army RAH-66

The TRADOC Systems Manager is responsible for the Operational Requirements Document and is the spokesperson for the PM on operational-suitability matters. For the Comanche, this individual is a colonel, assigned out of Fort Rucker, Alabama. A key organizational concept within the Army is the use of the TSM to represent the user community to the materiel developer.

A small TSM team is stationed full-time at Sikorsky and assists the PM and TSM in resolving operational-suitability issues and in holding discussions, and provides monthly reports to the PM and/or TSM. The operational requirements for the RAH-66 are stable: We were told that there have been *no* ORD changes since Boeing Sikorsky was selected to be the prime contractor team and award of the airframe contract (April 1991).

## DPRO SUPPORT HAS BEEN FIRMLY ESTABLISHED

The DPROs for all three programs are actively engaged with and used by all three programs' PMs in day-to-day management. We were told by both the PMs and their DPRO program integrators that the DPROs are fully integrated and are effectively supporting the programs, as is Defense Contract Management Command (DCMC), the DPROs' higher headquarters. All three programs have Memoranda of Agreements and program support plans signed by the respective PM and the Commander, DCMC.

Each of the DPROs has a PI assigned as a focal point at each facility. That PI works full time on the particular program. A lead PI, located at the lead or prime contractor facility, has a team of full- and/or part-time DPRO personnel to coordinate DPRO program activities at all locations and to support the program. DPRO members are members of various government and contractor IPTs, GDTs, and/or PDTs. Lead PIs are active members of PM teams and members of or support team personnel on Award Fee Boards. The DPROs also provide written reports, generally monthly, to their supported PMs.

## Navy F/A-18E/F

DPROs at all three contractor facilities (MDA, GE, and NGC) are actively involved in the day-to-day management of the program, ensuring that lessons learned from past programs are being applied to current programs and that the MOAs and program surveillance plans developed[32] for the F/A-18E/F are being followed. We were told that this program has priority for personnel resources at DPRO MDA (St. Louis, Mo.). The DPRO is active in contractor and government meetings and is definitely not treated as an outsider or sideline participant.

In fulfilling its program responsibilities, the DPRO communicates directly with PMA-265, as well as with its own parent headquarters, DCMC. Different chains of command (to DCMC from the Service Materiel Command) have apparently not resulted in confusion or failure in communication. At DPRO St. Louis, the MOA and surveillance plan call for a U.S. Navy program integrator to represent, act for, and coordinate efforts involving both DPRO and PM organizations. The PI, we were told, is being fully utilized, instead of being pushed off to the side.

---

[32]"MOA Between F/A-18 Program Office (PMA-265) and DPRO MDA," January 14, 1992, and draft update of January 14, 1992, MOA, July 22, 1994; and "DPRO, MDA F/A-18E/F Program Support Team Surveillance Plan," October 15, 1992, and draft update of October 15, 1992, Surveillance Plan, December 29, 1994.

## Air Force F-22

The DPROs at all four major F-22 team contractor facilities—LMAS (Georgia), LMTAS (Texas), Boeing (Washington), and P&W (Florida)—are actively engaged in day-to-day management of the F-22 Program. Operating under a program support plan,[33] each DPRO has a PI assigned full-time at each facility under the lead of the DPRO LMAS (Georgia) PI. Each PI has a team of full- and/or part-time DPRO personnel to support F-22 activities.

DPRO members are IPT members on each tier of all IPTs at the contractors' facilities. *No* SPO personnel are assigned at any of the contractors' facilities. The lead PI is an active participant in SPD meetings, a member of the Tier 1 IPT, and a member of the Award Fee Board. By being involved in the various IPTs, the DPRO team members contribute to the award-fee evaluation process. In addition, the lead PI provides a monthly 9-page assessment report to the SPD and the DCMC that includes program information from the four major DPROs.

## Army RAH-66

The DPROs are integrated in the management of the RAH-66 Program. Operating under PM or DCMC MOAs and surveillance plans, the DPROs at Sikorsky Aircraft and Boeing Helicopters have appointed a PI at each facility to coordinate and lead activities relating to the Comanche Program. Both PIs, in turn, report to a lead PI located at the Boeing Sikorsky JPO in Trumbull, Connecticut. The lead PI integrates all data from Boeing and Sikorsky and interacts with the Army PMO in St. Louis.

The PMO utilizes the DPROs both for oversight and as on-site representatives to help resolve program issues. The two DPRO PIs write monthly assessment reports to the PM through the lead PI. The DPROs' PI-led program support teams also input evaluations to the award-fee process on the Boeing Sikorsky airframe contract. The network of government PIs that has been set up at major program

---

[33]USAF, "F-22 Program Support Plan," approved by F-22 SPD and Commander, DCMC (date unknown).

subcontractor plants and facilities coordinates status reporting and enhances communication on program matters. To assess on-site process and system issues, the DPROs have established a joint management action team among the JPO PI, SA PI, BH PI, and the JPO Deputy PM. These PIs hold monthly meetings to discuss and resolve issues.

## INCENTIVES ARE APPARENT

Incentives can take many forms. The *Random House College Dictionary*, revised edition, 1980, defines the word *incentive* as "something that incites to action; stimulating; provocative; setting the tone" and calls it a synonym to the words *spur, incitement,* and *encouragement.* We are using the word *incentive* to describe the DoD environment/processes, attitude of senior DoD/Service management personnel, and actions of all associated government and contractor personnel and organizations that assist in the planning and execution of the particular acquisition program. We have tried to keep this discussion at the top level, rather than going into detail on pay/bonus systems for individuals and other means of recognition. Also, to some, dollars (profit) can be an incentive. Certainly, in one context, the award-fee provisions that allow the contractors to earn a profit on the basis of their performance is an incentive for them to achieve performance, schedule, and cost. In looking at where incentives exist, the authors have tried to go beyond this narrow, dollar focus.

To us, program success is the key incentive for these three programs, followed by knowledge that the users, or operators, have been provided new, modern, technologically superior weapon systems. The senior Service program management teams are clearly motivated to achieve these goals. Also, the fact that senior, experienced military officers are the PMs and SPDs means that they are focused on being role models for other, younger, newer PMs and on leading their Service's premier acquisition organizations.

A disincentive for the PMs and SPDs is the funding instability that affects each of the three programs and causes major perturbations to the programs (see the following section, "Funding Is Stable"). This disincentive notwithstanding, meeting performance, schedule, and cost constraints drives the programs and is the central focus of each program's reporting required by higher management. Incentives for each of the three programs are described briefly below.

## Navy F/A-18E/F

Management personnel on the F/A-18E/F Program are motivated primarily by the importance of this program for the Navy. It is Naval Aviation's number-one-priority program and is considered essential to the future of carrier-based tactical air warfare. PEO, PM, and NAVAIR are committed to learning from past programs. A business-as-usual attitude is proscribed. Innovative techniques are being sought and used. Lessons learned from past program mistakes are being studied to ensure that the F/A-18E/F does not encounter the same problems. Skills of individuals at all levels are being utilized.

## Air Force F-22

Incentives comparable to those in the F/A-18E/F Program are apparent in the F-22 Program:

- The F-22 is the USAF's number-one-priority development program.
- The SPD has two primary objectives:[34]
  - To develop and field the next-generation air-superiority fighter
  - To establish the standard for acquisition excellence.
- The SPD also realizes that E&MD costs and production costs are important and that they must be affordable if the program is to be a success.
- The LMAS Team PM expresses the two key incentives for the contractors' team as being[35]

---

[34]USAF, "F-22 SPO Briefing to RAND," WPAFB, Ohio, July 25, 1995.
[35]LMAS, "LMAS Briefing to RAND," Marietta, Ga., August 8, 1995.

— company (team) and individual integrity to meet program objectives and satisfy the customer

— profit (dollars earned from the award-fee process).

## Army RAH-66

Relevant aspects of program incentives include the following:

- The RAH-66 Comanche is the U.S. Army's number-one-priority acquisition program.
- The Army Chief of Staff provides strong personal support.
- The program has well-defined and stable requirements.

## FUNDING IS STABLE

Given the two most salient aspects of these programs—each is its Service's top-priority development program and each represents the future, with technologically advanced systems to be fielded in the post-2000 era—it would appear that the government would take any opportunity to achieve program stability, whether in its Planning, Programming and Budget System (PPBS) or in the congressional budget process. On the contrary. Budget instability plagues all three programs and causes the greatest concern for acquisition-management officials.

Failure to meet this criterion was the most seriously detrimental aspect we found during our research on the three programs. Where funding is concerned, a similar political and cultural environment pervades the Services, DoD, and congressional levels: Financial managers and leadership can and do, on an annual basis, change program funding levels, causing serious disruptions and reopening of major contracts, in a sole-source environment, in which a single contractor or contractor team is being dealt with in a noncompetitive process, to renegotiate and rephase efforts so that a program will function within the appropriations that are set annually.

## Navy F/A-18E/F

The burden of a Navy PM is increased severely when the Navy Comptroller (NAVCOMPT) makes fair-share funding cuts and separate funding reductions to a top-priority program after a major-milestone DAB review has been held—and independent of senior acquisition leadership approval. Rather than managing a program, the PM spends an inordinate amount of time justifying funding requirements and trying to regain budget and funding marks.

Figure 4.10 shows the history of adjustments to funding for the FA-18E/F Program since its DAB in May 1992. Program-funding levels have changed each year since the DAB, requiring adjustments to contractor work efforts through SOW changes and schedule sequencing, and to government oversight support and in-house (Navy) testing, and causing the Program Manager to be preoccupied with seeking recourse from top Navy leadership, taking time from actually managing the day-to-day execution of the program. Such adjustments are a regular occurrence within the Navy and are

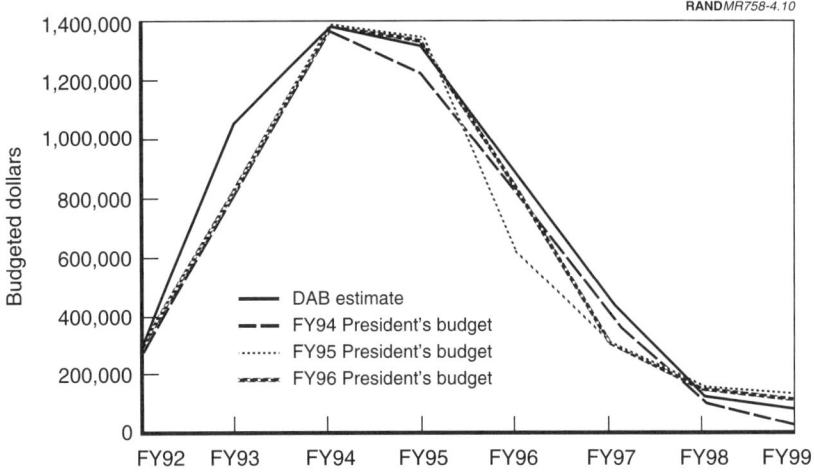

SOURCE: USN, "F/A-18E/F PMA-265 Data to RAND," Arlington, Va., February 1995.

Figure 4.10—F/A-18E/F E&MD Budget Chronology

thought of, generally, as a routine process for all programs to be faced with continuously.

Preoccupation with budget defense, usually against a claim of "schedule slippage," forces the PM to accord to schedule the highest priority in order to retain the program budget, rather than meeting technical performance objectives. We sensed this focus in the F/A-18E/F Program's priority to achieve first flight in December 1995 (its actual first flight was at the end of November 1995) or shortly thereafter to make the LRIP milestone and retain planned procurement budgets. The fear of losing funding support is a disincentive to proceeding in an orderly manner.

## Air Force F-22

Budget instability is a major problem for this program. Both OSD and Congress have made E&MD funding cuts since the Milestone II decision in 1991. The magnitude of these cuts and their effect on schedule and cost are shown in Figure 4.11. Each of the three program rephasings caused by funding reductions at the OSD and congressional levels required the government to reopen the contracts with LMAS and P&W and to rephase efforts to meet the new funding constraints. As a result, the schedule for the first flight of the first E&MD test vehicle slipped 22 months, and the Milestone III production decision slipped 32 months. Also shown are RDT&E-increase-induced higher negotiated costs of the rephasings, as well as the total program production-cost increases caused by the slippage of time and inflation.

The SPD's estimate of the consequences of such funding instability includes the following:

- Program stretched to 11+ years of E&MD
- Increased total program cost (E&MD and production)
- Increased damage to integrated product development
- Significant non-value-adding effort expended for each program and contract rephase
  — Design IPTs are either designing (CDR to first flight) or are engaged in rephase proposal builds

58  Three Programs and Ten Criteria

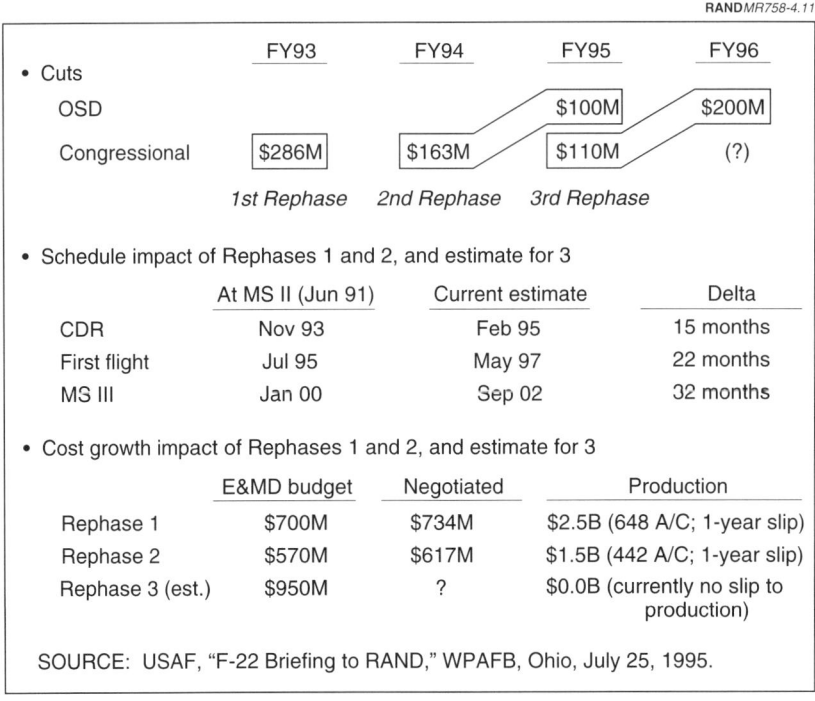

Figure 4.11—F-22 Cumulative OSD and Congressional Funding Cuts

- Subcontractor confidence increasingly eroded
- Rephase of program needed annually.

A model program must have maximum program (funding) stability. Funding cuts undermine the SPD's major objectives. While the Air Force has remained strong in its support for the program, support outside the Air Force has wavered.

## Army RAH-66

Army and DoD affordability considerations have required the RAH-66 Program to be restructured and streamlined. As a result, the scope of the program has been limited to two flying prototypes and six early operational capability (EOC) aircraft, and the acquisition schedule has been stretched out substantially, incurring cost increases (due to inflation effects). Figure 4.12 graphically displays the RDT&E funding perturbations to the annual program-funding levels that have occurred each year since 1991, as well as the consequences of the various restructuring and streamlining efforts.

Revisions to the competitively selected Boeing Sikorsky contract have had to be made in a sole-source environment. To defend the program and retain needed support, the PM has had to maintain a strong presence in the Washington area, which means that time has been diverted from managing the program to defending the program at high levels of the DoD.

## MANAGEMENT TEAM IS SELECTED FOR CREDIBILITY AND STABILITY

Ten or more years ago, it was common practice in the military departments to appoint Program Managers on the basis of their operational backgrounds, send them to DSMC for five months, then give them a major program to manage. Little attention was given to past acquisition experience as a prerequisite for PM positions. The Defense Acquisition Workforce Improvement Act (DAWIA) caused DoD to rethink its PM-selection process and to develop a more structured selection process. The DAWIA and DoD policy have set minimum education and acquisition experience levels for PMs.

In addition, it was commonplace to have PMs stay in their jobs only long enough to get a better job. Program-management continuity through a program phase or between formal milestones was a secondary consideration. Law and policy now set tour lengths for PMs.

As the senior acquisition officials in the Services, the Service Acquisition Executives have the approval/disapproval authority for major programs within their Service. Both the Army and the Navy convene formal boards for nominating candidates for PM positions to their SAEs for approval. Current Air Force practice is more informal: senior military leadership for acquisition meet with the Air Force SAE to select the individual. All three Services are formally documenting their processes, which include consideration of both military and civilian candidates and comply with the DAWIA and

60  Three Programs and Ten Criteria

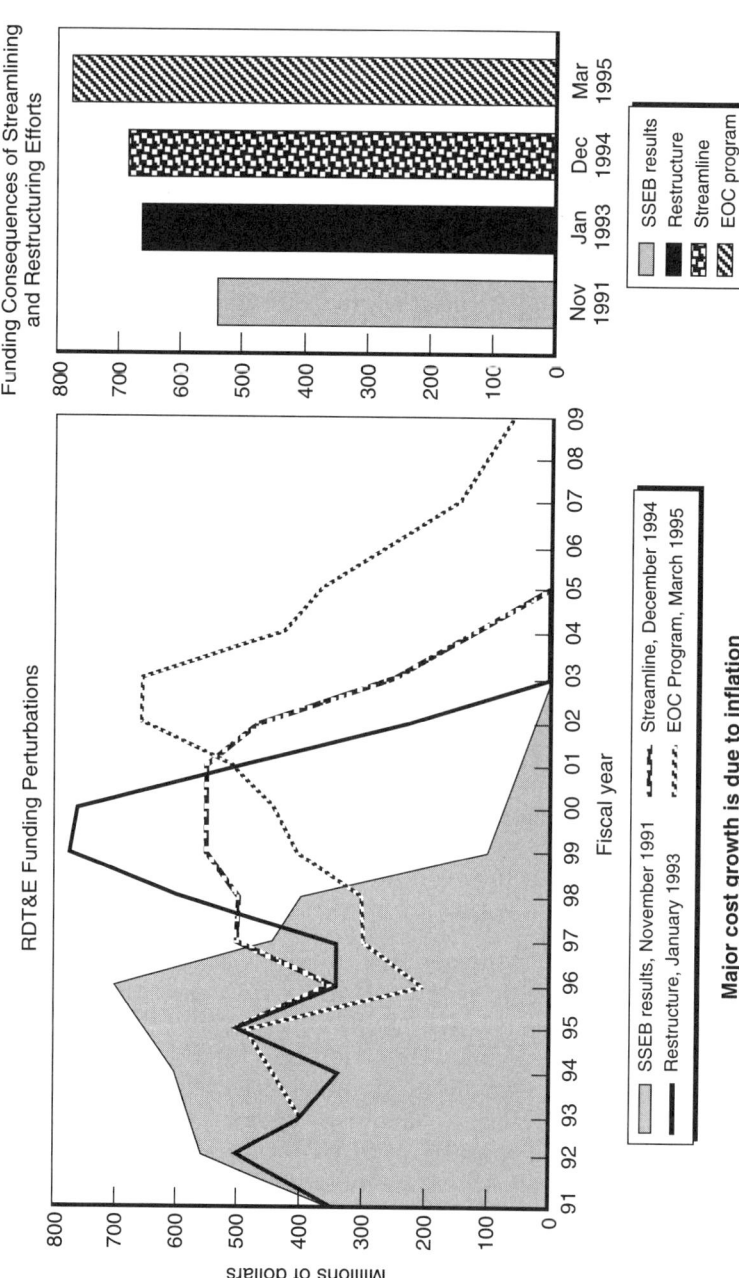

Figure 4.12—RAH-66 RDT&E Funding Changes

with the Deputy Under Secretary of Defense (Acquisition Reform) memo of May 23, 1994, "Assignment of the Best Qualified Individuals to Certain Senior Acquisition Positions," which specifies the requirement to develop and document (in writing) procedures for selection of PMs, considering both military and civilian candidates.

## Navy F/A-18E/F

To select and rank candidates for Program Manager positions, the Naval Aviation Program Executive Officers use a formal, objective documented process that is similar to that of the Navy military selection and promotion boards. DAWIA requirements are the cornerstone of qualification for each position. The flag officer, general officer, and civilian Senior Executive Service (SES) membership review official personnel records on previous performance in key acquisition and operational positions, at the Navy Annex, and conduct independent, secret voting. Final approval for the Navy is made through the Navy's Acquisition Workforce Oversight Council (AWOC) process, which is chaired by the ASN (RDA), who makes the final decision. With this approach, the Naval Aviation PEOs believe that personal biases will be avoided and that the best candidates will be selected for these important positions.

## Air Force F-22

Since DAWIA, the Air Force has maintained an informal process to ensure that qualified senior military officers and civilians are considered and nominated for major ACAT program-management positions. Procedures differ for general-officer/SES level and military officer (0–6 level)/GM-15 positions. The major commands in the Air Force maintain lists of both qualified military-officer and civilian candidates and provide nomination packages, when necessary, to the SAE for approval through the Director, Acquisition Career Management, on the Staff of the SAF/AQ. General officer/SES nominations are handled through their respective "General Officer Matters" and "SES Matters" offices on the Air Staff, which provide support to the SAF/AQ. The Director, Acquisition Career Management, is in the final coordination phase of a memo for SAF/AQ, formalizing the Air Force process to be used.

## Army RAH-66

The Army has established a centralized process for selecting PMs under the Deputy Chief of Staff for Personnel, Army Personnel Command. PM vacancies, identified by the PEO, are validated by the General Officer Steering Committee, chaired by the Military Deputy to the Army Acquisition Executive (AAE), who also carries the title Director of Acquisition Career Management. The PM-selection process at the colonel and lieutenant colonel levels is considered formal, objective, and well-documented: A formal Board convenes to develop candidate lists. Meeting DAWIA criteria is critical in this process. Priority lists are developed and provided to the SAE for approval, through the Army Personnel Command. The PM-selection process for general officers is less formal and involves only four key players: the General Officer Management Office, the Military Deputy to the SAE, the SAE, and the Chief of Staff of the Army. The SAE makes the recommendation, and the Chief of Staff appoints.

The Comanche PM has always been designated a general-officer position. We were told that the Army is evolving toward a system for evaluating military and civilians for PM positions in the same deliberative process. In January 1996, the formal Board was to begin working on an integrated order-of-merit, or best-qualified, list for presentation to the SAE. The Army is planning to develop an Army Regulation (AR 70-XX) within the next year to document this new process.

## SECURITY PROMOTES MANAGEMENT INVOLVEMENT

Security requirements do not appear to adversely affect program management on any of the three programs. The programs are certainly not being treated covertly as special-access programs (SAPs), nor does using security as a cover or excuse appear to exist to preclude proper and sufficient oversight of the programs. Special security needs are treated separately and include only those areas that are restricted. We were told that, in each of the programs, the PM is trying to downgrade security classifications of portions of the program whenever possible.

## Navy F/A-18E/F

We saw no negative effects on program management or oversight caused by security. On the contrary, we were informed that the current security guide was being reviewed and revised for downgrading, where possible, the classification of portions of the program.

## Air Force F-22

Security does not limit active program management. Special care has been taken to ensure that security does not become a reason for less-than-adequate management. Special security needs are treated separately; they include only those areas restricted by technological advancements of the F-22. The SPO continuously strives to reduce any special security restrictions placed on the program.

## Army RAH-66

Security classifications appear to have no adverse effects on adequate program management. The Comanche development effort is purposely limited to the SECRET level. Prior to Dem/Val, the low-observable (LO) aspects of the program were LIMDIS; however, this classification has been removed. The current level of classification ensures that technical and cost information will be available to all levels of management. The PM has developed an event-driven security classification guide and has a goal of all hardware being unclassified at the time of fielding.

Chapter Five

# SUMMARY OF OBSERVATIONS AND RECOMMENDATIONS

## SUMMARY OF OBSERVATIONS

Below is a summary of major observations the authors derived from this research effort. The list is not all-inclusive; it represents key points that address past program problems or lessons learned on issues raised in the public domain on previous DoD acquisition programs and addressed by senior program officials on these three DoD programs.

- The acquisition program responsibility, accountability, and reporting requirements of the Defense Management Review of 1989 are being followed.

- Requirements have stabilized. Progressive and continuous changing of ORDs has ceased. DMR has helped in this area.

- Qualified and experienced personnel are being selected and assigned to key PEO and PM positions. The selection processes are becoming more formal. DAWIA criteria are being used.

- PMs are being given life-cycle responsibility for their systems, even when those systems cross organizational and command lines. Charters including this responsibility are now being updated or written.

- Lessons learned from past programs are being taken seriously. There is much more communication among Service PMs and their staffs, and among programs across Service lines.

- Open communication within the Service acquisition communities is improving; a "no secrets philosophy" is an ultimate goal within each Service program. This improvement carries to the government/contractor link too, as well as filtering down to subcontractors.
- Use of IPTs (under various names) is becoming more prevalent in both government and contractor activities. IPTs are being given responsibility (in many cases) for monitoring technical performance, schedule, and allocated cost of their products. This is a major cultural change from past practices and will take time to mature.
- Contract performance measurements (TPMs, CS2, etc.) are important management tools and techniques and are being used, more and more, by both government and contractor managers as part of their management process, not just as a Contract Data Requirements List reporting requirement.
- Defense Plant Representative Offices (in addition to their DCMC responsibilities) are being used in program management as part of the PMs' teams.
- The Services consider the three subject programs as their top priority. However, serious disincentives face each PM, primarily in the form of budget instabilities from external perturbations and the lack of understanding of the consequences of the actions (e.g., budget reductions) being taken.

## RECOMMENDATIONS

Below are four recommendations that represent what the authors believe are the key outcomes of this study. The first three highlight the most important aspects of DoD program management at this time: achieving program stability, expanding the use of IPTs within DoD, and opening the communication channels within the government and between the government and its industry counterpart. The latter two subjects address key areas in which DoD is now attempting to change management processes to help improve program management.

- DoD must take action to stabilize the budgets for executing major, high-priority development programs in the Services. With current fiscal constraints, it might not be possible to protect every major program, but it should be feasible to protect the budget of one or two programs in each Service so that, after formal milestone review and approval, the budget could be changed only by the Service secretary and the military Service chief. Mid-level managers and staffs should be prevented from tinkering. Until something is done, DoD will, in all likelihood, continue to receive bad marks for program management because of major factors that are beyond the control of the Program Manager and his/her immediate superior, the PEO.

- DoD should support the evolution and maturity of IPTs within DoD and within industry as a good way to bring together multidisciplinary teams to work on a program; learn what is being done, good and bad; and share this information. The concept should be permitted to evolve, not be dictated from high levels of DoD. Both government and industry would benefit.

- DoD is now supporting open communication (no secrets; different media and forms) of real-time status. This support should be expanded, and the reporting of bad news should be encouraged by not taking immediate negative actions (such as automatically reducing the budget, creating outside-the-program special review teams to investigate the problem, or calling for a major program review by the milestone decision authority). The Services, PEOs, and PMs should be given time to analyze the situation and develop alternatives and recovery paths.

- As a valuable extension of this research and to compare how industry and the commercial world manage and operate, it would be helpful to DoD to assess similar, major commercial programs using the approach taken here. Comparing styles of management, processes used, incentives, and oversight techniques could give DoD useful information and insights.

Appendix

# OTHER SPECIFIC PROGRAM/SERVICE INITIATIVES

Besides the items discussed in Chapter Four, our research uncovered a number of practices that, to us, were unique to a particular Service, were unique in the way they were being accomplished, or were noteworthy due to the amount of effort being expended on them by the program. We have, therefore, decided to briefly discuss them in this appendix and trust that they will be of value to some program, in some Service, and that that program may want to contact the particular program—i.e., the F/A-18E/F, F-22, or RAH-66—for additional information.

## F/A-18E/F/NAVY

### System Engineering

A formal process for system engineering has been established by the airframe prime contractor. McDonnell Douglas Aircraft (MDA) has documented the procedures, which have flowed down through the project team. This process is apparently a "first" for MDA, and is being used for the first time on the F/A-18E/F.[1]

### Program Independent Analysis (PIA)

The program also uses a management tool called program independent analysis (PIA), in which coordinated groups of government,

---

[1] McDonnell Douglas Aircraft (MDA), "MDA Systems Engineering Implementation Briefing to RAND," St. Louis, Mo., February 1, 1995.

MDA, General Electric (GE), and Northrop Grumman Corporation (NGC) teams—independent and separate from their F/A-18E/F Program teams, who are chartered to perform independent "checks-and-balances" investigations and assessments of *particular* aspects of the program[2]—have an open charter to investigate *any* area of the program they deem appropriate or that has been identified by either the Navy or contractor PMs for analysis. The four independent teams meet monthly to review their ongoing activities and to coordinate future actions.

Some examples of recently completed PIAs include

- preparations for the first-flight readiness review
- development of full-authority digital engine-control software and change process
- preparation of the interactive electronic technical manuals.

Upcoming and ongoing PIAs will look at, among other things, fleet supportability of composite materials, preparations for the operational test readiness review, the effect of integrated product team (IPT)/contract administrative officer (CAO) reorganization on all flight test teams, and Engineering and Manufacturing Development (E&MD) engine supportability. This F/A-18E/F PIA activity is patterned along the lines of the Navy's Strategic System Programs Office (SSPO) concept, in which an independent section in their government program office was used for conducting similar off-line analyses.

## F-22/AIR FORCE

### Integrated Product Teams

The use of integrated product teams on the F-22 is the key to the management approach of the program and was a major undertaking

---

[2]"PMA-265 Program Independent Analysis (PIA) Team Handbook," October 26, 1992; "PMA-265 Program Independent Analysis Overview and Guide," December 6, 1994; MDA, "Establishment of the Office of F/A-18 Program Independent Analysis Team," St. Louis, Mo.: MDA Memo, undated; MDA, "MDA Program Independent Analysis Briefing [to RAND]," St. Louis, Mo., February 1, 1995.

on the part of the Air Force and the contractor team. It required a significant cultural change from function-oriented organizations to product-oriented team organizations, the evolution of which is depicted in Figures A.1 and A.2.[3] The two resulting organizations (one for the Air Force and one for the contractor) also mirror each other (deliberately), as shown in Figure A.3.

Organizational changes in themselves do not guarantee success, but the approach taken by the System Program Office (SPO) Program Director (SPD) has been to assign the IPTs the responsibility and authority for their product's performance, schedule, and cost (including risk management). This product responsibility is referred to by the SPD as the Iron Triangle (because of its performance, schedule, and cost "legs"). It is through these IPTs that the program is being executed. The lesson learned from the SPO with respect to IPTs, we were told, is that people are the key. Having the right people is important; if they cannot work in a teaming arrangement, they must be replaced. Care must also be taken to ensure that the IPT's responsibility does not become overbearing for the higher-level IPT. For this reason, each tier of the F-22 IPTs has an Analysis and Integration IPT to help that higher-level IPT integrate all aspects of the next lower-level set of IPTs. Both the SPD and the Lockheed Martin Aeronautical System (LMAS) F-22 team leader judge that the IPT-oriented organizations have worked well.

## Integrated Master Plan (IMP)/Integrated Master Schedule (IMS)

The Air Force acquisition approach is to use the concept of integrated master plan and integrated master schedule (IMP/IMS) to plan and execute a program. This concept is the basis for managing against accomplishments and their related exit criteria of what must be satisfied prior to successful completion of a particular event or activity. The F-22 Program uses this *event-based*—i.e., based on the Statement of Work—*IMP*—a contractual document—as the central means of managing what is to be accomplished. The IMP is the con-

---

[3]U.S. Air Force, "F-22 SPO Briefing to RAND," WPAFB, Ohio, July 25, 1995.

72   Three Programs and Ten Criteria

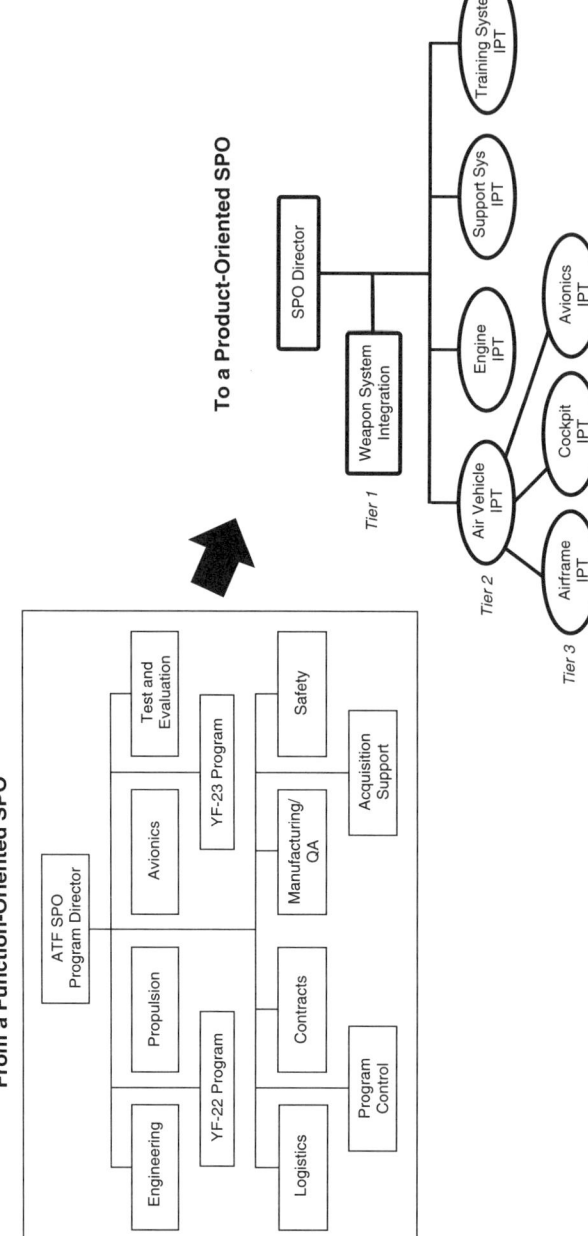

Figure A.1—F-22 SPO Organization Changes

SOURCE: USAF, "F-22 SPO Briefing to RAND," WPAFB, Ohio, July 25, 1995.

Other Specific Program/Service Initiatives 73

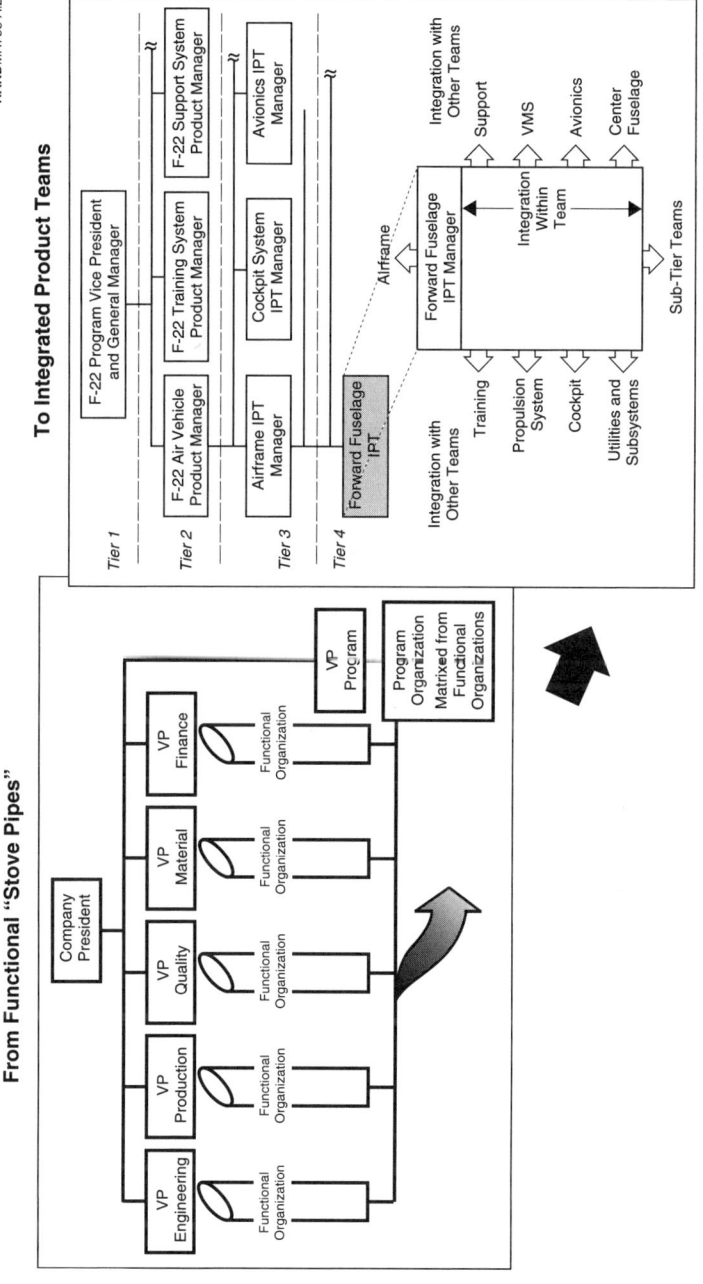

Figure A.2—LMAS F-22 Organization Changes

SOURCE: USAF, "F-22 SPO Briefing to RAND," WPAFB, Ohio, July 25, 1995.

74    Three Programs and Ten Criteria

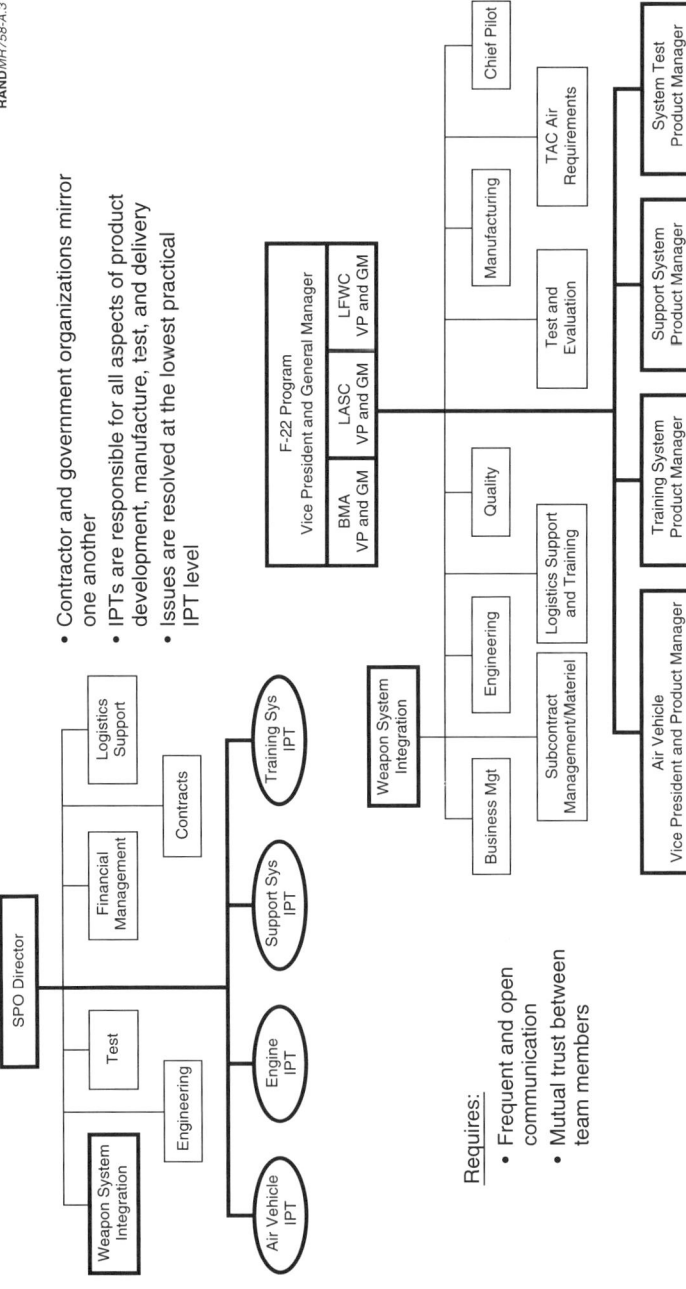

Figure A.3—SPO/LMAS Organizational Compatibility

SOURCE: USAF, "F-22 SPO Briefing to RAND," WPAFB, Ohio, July 25, 1995.

tractor's plan to design, develop, test, and deliver the F-22 development aircraft, and the efforts associated with those approximately 13,000 F-22 events or activities identified in the F-22 air-vehicle contract. The IMS places the events, criteria, and accomplishments against a timeline; some 30,000 activities are associated with the IMS. The IMS is not contractually binding (it is a Contract Data Requirements List requirement). The SPO uses the IMP and IMS together to execute and track the program.

## Requirements Traceability Matrix (RTM)/System Maturity Matrix(SMM)

These two very important processes are used on the program to track the operational requirements as they flow down from top-level system specifications to lower-level hardware and software specifications. In execution, the system maturity matrix tracks the progress—the increasing maturity—of a requirement through its E&MD phase, by the activities that are accomplished, against the timeline plan of the technical performance measures that have been established. Both of these processes are managed by the IPTs.

## Lean Enterprises

The F-22 SPO/contractor team realized early in the E&MD phase that affordability was key to the future of the F-22. They developed a process involving both the primes and key subcontractors and having the key objectives of employing the best practices of industry, improving manufacturing efficiency, and minimizing cost increases resulting from rate and quantity changes. According to the team, "whatever you measure will improve."

Areas being investigated and traced over time with metrics include production cost estimate, lead-time reduction, design changes, scrap, rework and repair process capability, inventory turns, incoming and source inspection reductions, overhead (cost) control, and supplier participation. Both airframe and engine primes, and a significant number of their major subcontractors and suppliers, are actively participating.

## Lightning Bolt Initiative

The Assistant Secretary of the Air Force (Acquisition) (SAF/AQ) has undertaken an intense initiative to make "bold sweeping changes" in the way the Air Force runs its acquisition programs.[4] Recently initiated (spring 1995), the initiatives (nine to date) are aimed at implementing change in a relatively short time frame (most less than six months to develop, implement, and make operational). Key is that an individual (not a committee) is responsible for implementation. Of the nine initiatives so far, one is the responsibility of the Material Command, five are led by SAF/AQX (Deputy Assistant Secretary of the Air Force [Management Policy and Program Integration] of SAF), and three are led by SAF/AQC (Deputy Assistant Secretary of the Air Force [Contracting] of SAF) individuals. These nine are as follows:

- Establish a centralized Request for Proposal (RFP) support team to scrub all RFPs, contract options, and contract modifications over $10 million.

- Create a standing Acquisition Strategy Panel composed of senior-level acquisition personnel from SAF/AQ, Air Force Material Command (AFMC), and the users.

- Develop a new SPO manpower model that uses the tenets established in the management of classified/special-access program (SAP)-level programs.

- Cancel all AFMC-center-level acquisition policies by December 1, 1995.

- Reinvent the Air Force System Acquisition Review Council process.

- Enhance the role of past performance in source selection.

- Replace acquisition documents with the Single Acquisition Management Plan (SAMP).

---

[4]SAF/AQ, "Air Force Lightning Bolt Initiative," from the Office of the Assistant Secretary of the Air Force (Acquisition), undated, and Update No. 3, dated July 20, 1995.

- Revise the Program Executive Officer (PEO) and Designated Acquisition Commander (DAC) portfolio review to add a section that deals specifically with acquisition reform.

- Enhance Air Force acquisition workforce with a comprehensive education and training program that integrates acquisition-reform initiatives.

## RAH-66 COMANCHE/ARMY

### Design Flexibility

In executing the Comanche Program, Boeing Sikorsky has been given design flexibility to tailor any of 19 different requirements within the weapon-system specification (contractual specification meeting the Operational Requirements Document [ORD]), within established limits for optimizing the weapon system's design.[5] Three categories of changes or tailoring have been set up: (1) those affecting one or more design-flexibility parameters (within the specified bands of each parameter); (2) changes outside the acceptable bands of that parameter or that will change significant system-level attributes, safety, or exit criteria; and (3) administrative changes or other minor descriptive changes to the program. The Army changed its approval process for categories (1) and (3) so that only disapproval is given within five days or the change is automatically approved. For category (2), the government has 15 days to approve or disapprove the proposed changes. In all categories, *every* change must remain in full conformance to the ORD. Since source selection, 81 changes have been approved and incorporated throughout the process. We were told that the majority of changes to date have contributed to both cost avoidance and weight reduction.

### Business Planning

Under the heading of "business planning," the Program Manager and Aviation and Troop Command (ATCOM) have established a process that annually develops a comprehensive business plan detailing negotiated five-year support plans with each interfacing

---

[5]U.S. Army, "RAH-66 PMO Briefing to RAND," St. Louis, Mo., August 16, 1995.

government organization supporting the program. This plan provides an understanding of and visibility into which organization will be doing tasks in support of the PM. It also provides the PM substantiation to help budget for and obtain the funds necessary to reimburse those organizations. Through this process, the PM also gets the commitment of these organizations to provide the number and mix of skills required to support him or her. Holding the funding for these organizations' support also gives the PM leverage to ensure satisfactory and quality efforts on his or her behalf.

## Team Comanche[6]

In 1991, the PM, PEO Aviation, and Commander of ATCOM established a formal process for problem resolution. Referred to as Team Comanche, this process has three levels: the process-action team (PAT), the management working group (MWG), and the Executive Steering Group (ESG).

The function of the PAT is to identify and resolve problems at the lowest-level, earliest-possible stage. Membership consists of representatives from the Program Manager's office, ATCOM, other major subordinate commands (MSCs) as necessary, Training and Doctrine Command (TRADOC), DPRO, and the prime contractors.

Issues that cannot be resolved are raised to the MWG, which consists of the PM, the Boeing Sikorsky Joint Program Office, the engine contractor PM, the TRADOC Systems Manager (TSM), and the Executive Director of the Aviation Research, Development and Engineering Center. Issues that have major programmatic impact and/or that cannot be resolved by the MWG are brought to the ESG. The ESG comprises the Military Deputy to the Assistant Secretary of the Army for Research, Development and Acquisition (ASA [RDA]); the PEO; the commanders of ATCOM, Communications and Electronics Command (CECOM), and the Army Aviation Center (Ft. Rucker, Alabama); the airframe and engine companies' presidents; and the Deputy Under Secretary of the Army for Operations Research.

---

[6]U.S. Army, "RAH-66 PMO Briefing to RAND," St. Louis, Mo., August 16, 1995.

We were told that this process has worked well and is credited with materially assisting the restructuring and streamlining initiatives, and getting them through the Army leadership in an efficient and timely manner.